Tomorrow Has Wings

The Story of Captain Jack Alcock
First to Fly the Atlantic Non-Stop in 1919

Joseph F Alcock

Aerospace Publishers

PALM BEACH GARDENS, FLORIDA

Tomorrow Has Wings: The Story of Captain Jack Alcock First to Fly the Atlantic Non-Stop in 1919
by Joseph F Alcock

First published by Aerospace Publishers in 2013
Palm Beach Gardens, FL
Revised 2025

ISBN: 978-0-9890111-0-5

Cover: Portrait of Captain Jack Alcock by Ambrose McEvoy, British portrait artist (1878-1927), from a photograph of the original painting owned by Tony Alcock.

Books by Joseph F Alcock:

- *Tomorrow Has Wings – The Story of Captain Jack Alcock, First to Fly the Atlantic Non-Stop in 1919*
- *Tomorrow Is Supersonic – The Story of the Jet Age*
- *Tomorrow the Moon – The Story of the Space Age*

But they that wait upon the Lord shall renew
their strength;
they shall mount up with wings as eagles…
Isaiah 40:31 (KJV)

Contents

Contents (Continued)

Author's Note

I have written this book about Sir John (Jack) Alcock to give the reader insight into his entire life, not just his historical flight across the Atlantic. The story is told by Jack himself and relates actual events he experienced. Much of the dialog is fictional, but it is based on conversations that did take place and, where possible, includes the actual words used by the characters in the story.

Some of the photographs are of lesser quality than desired due to the age and condition of the originals. I have made every effort to capture the language of his era, including the spelling and punctuation.

Joseph F Alcock

Preface

When I was in high school, my father told me about the first transatlantic flight, which took place before Lindbergh's famous flight. It was flown by two British aviators, Alcock and Brown (John William Alcock, KBE, DSC, RAF[1] and Arthur Whitten Brown, KBE, RAF). Later, when I was an aerospace engineer, my colleagues occasionally would ask if I were related to John Alcock. This piqued my curiosity and led me to trace both families to the 1700s. I found no connection, although both families were from the same region of England.

While researching John Alcock, I learned he had been an extraordinary aircraft mechanic and aviator, who had made major contributions to aviation in the early years, in addition to his famous flight across the Atlantic.

Jack Alcock, as he preferred to be called, began flying before World War I and was associated with some of the great names in early aviation During World War I, he participated in many air raids and reconnaissance missions in the eastern Mediterranean and became a prisoner of war in Turkey. During his captivity, he formulated a plan to be the first to fly across the Atlantic non-stop and win the Atlantic Prize offered by the *Daily Mail*, a London newspaper.

His story inspired me to go to England and meet two of Jack's nephews, Tony and Neville Alcock, and visit the Museum of Science and Industry in Manchester, where much of Jack's memorabilia, correspondences and related newspaper articles are archived.

Joseph F Alcock

Acknowledgments

I am very appreciative of the encouragement of the Abacoa Writers Group, and from my wife, Lois, and my daughter, Debra Alcock, John and Grace Tadler, and my former boss, Lawrence G Tanner.

I enjoyed meeting two nephews of Captain Jack Alcock: Group Captain Anthony J H Alcock, MBE, RAF[1] and W Neville Alcock. Tony and his wife, Rita, and Neville and his wife, Judy, were very cordial and very helpful during our visit with them in September 2011. They provided photos and information, some of which had never been published.

Nick Forder, Curator (Air and Space) at the Museum of Science and Industry (MOSI) Manchester, England, is an expert on the many aircraft of Jack Alcock's era. He provided a considerable amount of information on Jack Alcock and the many aircraft he was associated with, which greatly enhanced the story.

I'm deeply indebted to Collette Curry, of Manchester, England, a close friend of the family. Collette provided many of the photographs in the book and also supplied information on Jack Alcock's life. s. Curry is a PhD candidate at Manchester Metropolitan University, where she is researching and programming a reminiscence aid for older people with early dementia using artificial intelligence techniques.

I am very appreciative of John Pulford, Head of Collections and Interpretation at Brooklands Museum, for all his help and photographs relating to the early days of the Brooklands aerodrome.

Introduction

Prior to World War I, the airplane was in its infancy. The airplane structure, its engines, and its control system were very basic and, in some cases, very flawed to the extent they sometimes claimed the life of the pilots who flew them. The war brought the airplane to a level capable of a transatlantic flight. This certainly was the case of the Rolls-Royce powered Vickers Vimy, a World War I bomber, that Captain John William Alcock and Lieutenant Arthur Whitten Brown used in their historic non-stop flight across the Atlantic in June 1919.

This is somewhat of a surprise to some Americans who thought Charles Lindbergh was the first to cross the Atlantic non-stop in 1927. However, Lindbergh was the first *solo* non-stop flight across the Atlantic. Both flights were of historic significance. Alcock and Brown's flight was significant because it was the first; Lindbergh's because he flew solo and for a much greater distance.

Prologue

England, September 1919

Although the actual time occupied in my transatlantic flight was under sixteen hours, it may be fairly said that it took ten years to accomplish. It was perhaps owing to the knowledge of this fact that the public appreciation of the feat overwhelmed me in no small degree. To them it was a sudden and perhaps sensational achievement, but it was really the successful outcome of much preparation. There is always satisfaction in being the first to do anything, whatever it may be, and while the lapse of time may intensify that feeling, the flight was nothing more than the execution of a long-conceived plan.

Ten years of continuous flying undoubtedly stood me in good stead, and had I been less fit or less experienced, the excellent machine produced by Messrs. Vickers and the reliable Rolls-Royce engine might have been sacrificed. My story began many years before I took up flying.

Jack Alcock

Chapter 1 – Early Years

Manchester, England, 1900 -1908

WHEN I WAS SEVEN YEARS OF AGE, my father said to me, 'Johnny, this is the dawn of a new century, and it shall be exciting. The world is changing. The time is coming when there will be no horse-drawn carriages. Instead, horseless carriages shall be commonplace. They are being built in England, in Europe, and in America.

'Here in England, Charles Parsons built a machine that flew 300 feet. Steam expanding through a tube to the atmosphere pushed the craft in the opposite direction, and it was held aloft by its wings.'

Another time, my father spoke of internal combustion engines that were used in horseless carriages and motorbikes, and he spoke of steam engines used to power locomotives.

His conversations caused me to be intrigued by machines that could transport a man. My desire was to learn as much as I could about them. At primary school,[1] and at home, I often read articles and manuals on these machines.

In those days, there was little written about flying machines. The achievements of Orville and Wilbur Wright's flying machines were not well reported here and in Europe. Few people who had heard of their flights believed they had actually flown. This was the reason Wilbur Wright made several flights in France in 1908.

Several years later, I learnt the story of the Wright brothers' achievements.

The Wright brothers fitted a motor of their own construction to a glider and proved that the machine could rise from the ground with a pilot in control. They made four flights of 120 to 850 feet with their *Flyer* in December 1903.

In November and December of 1904, they flew their *Flyer II*, each time exceeding five minutes and traveling nearly three miles. By autumn 1905, they were flying distances up to 24 miles in their *Flyer III*.

The Wright brothers became more interested in flying machine control rather than larger engines, whilst most of the others seemed to place greater importance on larger engines.

I also learnt later that a Brazilian, Alberto Santo-Dumont, famous for his work with dirigibles and his flight around the Eiffel Tower in 1901, made the first successful flight in Europe in a winged machine in 1906.

At Manchester Central High School, I began signing my schoolbooks, Jack Alcock. I also requested my friends to call me Jack to distinguish me from my father, whose name also was John.

Before I left school in 1908, I built a balloon using bamboo and silk. I did not use helium or hydrogen to inflate it, but the construction gave me a feeling that aerial devices could be built if one sets one's mind to it. However, my sister, Elsie, and my younger brother, Edward, made sport of my early experiments. Interestingly, they treated me much better when I made Parkin[2] and treacle toffee, which they quickly devoured.

Just a few years later, I became very involved in aerial devices.

Wright *Flyer III*, October 1905.
(From National Register of Historic Places, ref. no. 90001747)

Chapter 2 - Beginnings

Manchester, Bolton and Longsight, England, 1908 - 1909

ONE DAY after I left Manchester Central High School in 1908, Father said to me, ‘Let us take a ride,’ as he motioned to a horse-drawn wagon used for circulating newspapers. Father being a very tall man, over six feet in height, easily stepped up into the wagon as I climbed into it. From the serious look on his face, I knew there was something important he wanted to discuss with me.

Father got right to the point. ‘Jack, I have been considering your future, and I would like to know what you would like to do to

Father and Mum: John and Mary (Whitelegg) Alcock
(From W. Neville Alcock)

make your living? I think you have made plans already. What are they?'

'I would like to become an engineer and eventually work on horseless carriages and motorbikes.'

Father was without expression. He simply said, 'You obviously have an interest in mechanics, which I recognise and support. I have an interest in these things, too.' He paused and pulled on one of the reins to guide the horse onto the main road; he continued, 'Mother and I shall help you however we can.'

I was elated. I would be able to do something I very much wanted to do. I thanked Father and Mum for their support.

<><><>

SHORTLY AFTER, I BEGAN WORKING at Empress Motor Works, on Stockport Road in Longsight. This was a large works about a mile east of where my family lived, but more importantly, it built and repaired motorcars and motor bicycles.

16-20 H.P. Empress Car
Empress letterhead from W.D. Korsdike letter, 3 December 1909 at Manchester Museum Of Science and Industry (MOSI, Ref. A1997.36/4).

I was assigned a toolbox and was given a work schedule. A master mechanic worked with me until I became more experienced. I soon became thoroughly at home in the workshops, and mechanics seemed a natural thing to me. Empress made six different motorcars for the upper middle class. Two were particularly popular, one with a 20 h.p., in-line, four-cylinder engine and a wheelbase of 114 inches, and another with a 30 h.p., in-line, six-cylinder engine and a wheelbase of 138 inches.

However, I soon made a name for myself, but not in a manner that I would have desired. When driving the works lorry[1], suddenly I reversed it into a workshop wall.

Mr. Korsdike, the owner of Empress at the time, heard about my mishap and quickly had Mr. Fletcher, his general manager, inform me, 'You are not permitted to operate the company vehicles until you demonstrate that you can drive properly.'

This was not the way I intended to begin my career.

<><><>

IN 1909, I BEGAN TO HEAR OF £1,000 prize money offered by the London newspaper, the *Daily Mail*, for the first successful flight across the Channel. The newspapers favoured Hubert Latham to win the prize. However, on July 19th 1909, he failed in his first attempt in his Lavavasseur *Antoinette IV*.

Six days later, on July 25th 1909, while at work, I saw all the workers milling around the works floor. Before I could ask about all the excitement, somebody came up to me and said, 'Jack, have you heard Louis Bleriot successfully flew from France to England to claim the £1,000 prize the *Daily Mail* promised to the first flyer who crossed the Channel?'

I became interested in his flight. Later, I read reports that said Bleriot was an inventor, an engineer, and a pilot. He was experienced in both building and flying machines. His machine[2] had a single wing, which he claimed was 'more aerodynamically efficient,' meaning it did not have as much drag as a two-wing machine.

However, the single wing did not have the rigidity of two wings wired together. The weaker structure of his machine made me wonder if this design would be successful. I later read about only one other single-wing machine. It was designed and built by a Romanian, and it flew only forty feet.

That night I returned home and told Father about Bleriot's flight.

Louis Bleriot
(From the Public Domain)

'Jack, his accomplishment raises some concern. Many in our government and military are very upset about it. Our leaders know that his feat means that Britain can be attacked from the air. Just read the London newspapers. However, an aerial attack from the mainland is not going to occur in the near future, because the flying machines now are quite limited. Many of the engines tend to overheat and seize-up after they have been running about twenty minutes. Bleriot was very fortunate in his 37-minute flight. The heavy rain he encountered at the end of his flight cooled his engine enough to allow him to stay aloft for as long as he did.'

Father paused and leaned forward in his chair to ensure that I could hear what he was about to say, 'Part of the future of aviation is in making the engines last longer and not overheat.'

He paused again and kept his voice barely above a whisper, 'Jack, some things happening in Germany are cause for alarm amongst our leaders. Bismarck conquered France 40 years ago, and now the

Kaiser appears to have similar military ambitions. He has been expanding the German navy so that it is now a major force in the North Sea. He appears to be unstable at times, which makes one wonder if he will start a war.'

'If British companies like Empress develop better engines, then we will be more prepared to defend ourselves,' I suggested.

At that point, Father became very philosophical and in a sad voice and expression said, 'Someday, everyone will be able to fly in machines that will go much farther than Bleriot's. It probably will be the result of a major war. It is a bad commentary on the human race that so many improvements are the result of technical achievements from warfare.'

1909 Poster of Louis Bleriot leaving France to cross the Channel
(From the Public Domain)

<><><>

DURING THE SUMMER of 1909, Mr. Fletcher began work on a monoplane. No doubt, he was inspired by Bleriot's recent accomplishment. By September 15th 1909, the design had been completed. When the construction was about to begin, he approached me and said, 'Jack, I like your work, and I would like you to help me build my machine, which I intend to take to a flying meeting at Blackpool.'

I was excited to work on his machine. I did whatever I could to get it completed correctly and in a timely manner, so he could advertise it at Blackpool, where many British and French aviators with their machines met.

The framework of Mr. Fletcher's machine was bamboo with the exception of the engine support, which was steel. The elevating and balancing planes were connected to pedals by wires, and a vertical wheel from a motorcar was connected to the rudder for steering. There was a new feature to withstand the shock of landing. Most aeroplanes, so far, have been fitted with springs, but Mr. Fletcher has developed cylindrical air cushions for this purpose. Another new feature was the wings were on a hinge and a bolt. When the bolt was withdrawn, the wings folded up, and the machine was easily carried.

In early October, one of the clerks at Empress came to me and said, 'Jack, there is a long package from William Cochrane. It probably is for Mr. Fletcher's machine.'

It was the propeller, which was made of steel tube laced across with wire and then covered with canvas. It was about six feet long and weighed less than seven pounds. Cochrane, when he visited Mr. Fletcher, claimed, 'My propeller has greater breaking strain than either wood or metal, and it gives more thrust.'

Sometime after 6:00 a.m. on October 13th 1909, we went to Heaton Park, which was a large area of 600 acres about five miles north of the centre of Manchester. It was a rainy morning, but Mr. Fletcher was anxious to try his machine. I tried to start the five-cylinder engine, but it would not go. We all had a turn at the engine without success, and we became very warm and wet in our attempts. Saddened by the situation, Mr. Fletcher housed the machine in a shed near Heaton Hall, where we worked on the engine and were able to get it to run.

<><><>

ON OCTOBER 18TH, I accompanied Mr. Fletcher at the 'Flying Meeting' at Blackpool. The first man to start was A.V. Roe who brought his small yellow machine out of its shed to test it. The machine refused to take off, and he lost the chance for an Englishman to be the first to fly at Blackpool. Thus, A.V. Roe gained the title 'The Hopper.' I was told later that he had the first successful flight of an all-British machine the previous July. The meeting was the highlight of my life at that point, as I saw many machines.

The French, in particular, had a major presence at the meeting. The first to fly was the Frenchman, Henri Farman, who finished a flight with applause from the 200,000 spectators. He also set the first official British record with a flight of just over 47 miles and then won the Grand Blackpool Prize of £2,000 for completing the greatest number of circuits without touching the ground.

Mr. Fletcher tried his machine again at Heaton Park on the evening of October 20th 1909. After some runs of about 100 yards along the ground, he was able to get the 365 lb. machine[3] off the ground for about 20 or 30 yards. One of the mechanics standing near me suddenly shouted, 'He has a problem!'

Mr. Fletcher was unable to control the machine when it was caught by a gust of wind, and it was thrown over on its side. Fortunately, he was not injured, but one of the wings was slightly damaged and three of the wheels had buckled.

Shortly after this incident, Mr. Fletcher started to build another machine similar to Antoinette and Bleriot monoplanes. He had viewed the details of these designs close up at Blackpool. These two machines not only influenced Mr. Fletcher, but they influenced the design of monoplanes in Britain until about 1911, when the Nieuport IV began to influence the designs of Hanriot and Avro monoplanes, and the design of French Morane and Deperdussin monoplanes.

Charles Fletcher's first monoplane.
(Courtesy of Michael H. Goodall)

Chapter 3 – Empress Aviation

Longsight and Manchester, England, 1909-1910

IN NOVEMBER 1909, I became 17 years of age, and at that time Charles Fletcher acquired the business from W.D. Korsdike. Mr. Fletcher made the company a public enterprise by issuing 15,000 shares of stock valued at £1 per share. He also gave company stock to Mr. Joseph Tonge, in appreciation for his service to the company. Mr. Fletcher was very interested in aviation and renamed the company, The Empress Motor Car and Aviation Co., Ltd. He also had a close relation with the Manchester Aero Club and one of its founders, Norman Crossland, a local engineer.

Mr. Fletcher now could turn more of the company's activities towards his interest in aviation and make several modifications to his monoplane, based on his prior experiences.

<><><>

IN DECEMBER 1909, Mr. Fletcher called the mechanics and me into his office and said, 'Ready the monoplane for some aerial testing at Heaton Park on the 29th.'

On that day, the weather conditions were unfavourable, causing the carburettor to freeze up, which prevented any flights. However, in January we attained some success. The machine, with an Empress engine, made several 75 to 100 yard flights.

On January 17th, in another test, just as Mr. Fletcher was about to take his seat in his machine, a mechanic shouted, 'Oh, blimey.'

With that exclamation, Mr. Fletcher backed away from the machine just as a gust of wind caught it. The machine lifted some 15 feet, collided violently with a clump of trees and suffered considerable damage to the wings and tail plane. It was eventually restored but received its final fate in March when it was put on display at the Manchester Aero Club Show in White City.

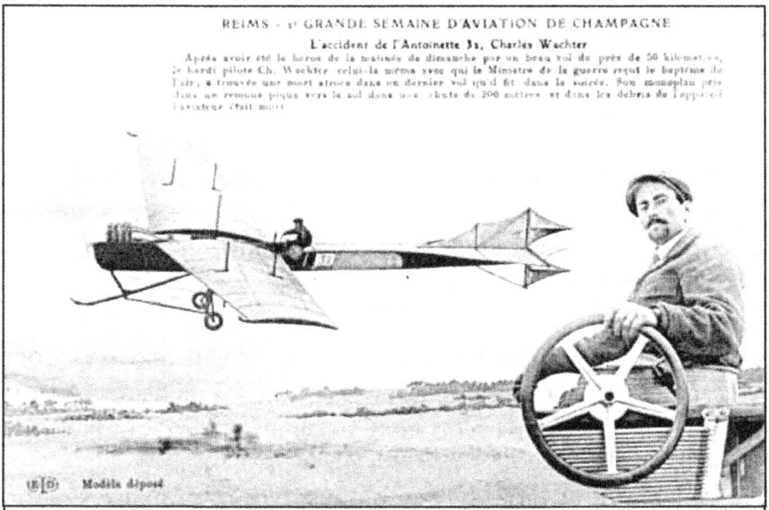

Antoinette VII, flown by Hubert Latham in his second attempt to cross the Channel on 27 July 1909. Fletcher's machine was based, in part, on the Antoinette design. Maximum speed: 44 mph. (From the Public Domain; pre-World War I postcard)

Mr. Fletcher's last attempt at building his own machine came in February. One of the wealthy members of the Manchester Aero Club, a local businessman, asked him to build a flying machine for him. In due course, with the help of Mr. Crossland, he set about constructing a biplane. The completed machine was impressive with its mahogany fuselage and bright brass work. In March 1910, it was tested at the Manchester Racecourse at Castle Irwell. It made a few erratic hops but never got off the ground!

Mr. Fletcher and the local businessman were greatly disappointed. However, I heard Mr. Crossland say, 'Do not give up, Charles, there is always something to be learned from our mistakes and failures.'

'I shall continue to develop my engines, but I need to think about building another machine,' Mr. Fletcher replied.

He continued to build engines and test them.

<><><>

JUST BEFORE I WENT TO WORK one day in April 1910, while taking my breakfast, I read the following article in one of the newspapers: Claude Grahame-White will attempt on April 23rd to claim the *Daily Mail* £10,000 prize for the first flight from London to Manchester within a 24 hour period.

I told Father of the contest. He could easily observe that I was excited about this upcoming event, which would require a successful flight of about 180 miles. 'Let us go to Manchester on the 24th and see him land. The landing will be at Barcicroft Fields, near Didsbury, just two miles from here,' he suggested.

Louis Paulhan, 1909.
(From the Public Domain)

We travelled to Manchester on our bicycles, but soon after we arrived, we learnt that Grahame-White's attempt had failed. After he landed, near Lichfield, to rest the previous night, high winds overturned and damaged his machine. He was just 50 miles from his goal. The following day, we read in the newspaper that he had returned to London to repair his machine. Soon after, the Frenchman, Louis Paulhan, having heard of

Grahame-White's misfortune, decided to attempt the flight. Thus, Paulhan's entry created a competition to Didsbury, which began on Wednesday, April 27th.

Upon hearing of Grahame-White's second attempt on the wireless, we realised that it had become an air race, and we were even more excited about seeing the finish. 'Let us go there once again,' Father said.

'Yes, of course, t would be capital should he, an Englishman, win the prize,' I replied.

My father and I bicycled to Didsbury, arriving before sunrise on Thursday to await the arrival of the two flyers. They had left the Lichfield field in the dark in order to finish their flights within the prescribed 24 hours, should they be delayed. Shortly before 5:30 a.m., we saw a Farman descending toward the clover covered field. Both Grahame-White and Paulhan were flying Farman machines, so we were unable to determine who was landing in the field.

'By Jove, it is the Frenchman,' my father exclaimed when he spotted the pork pie hat and white scarf that Paulhan was reported to be wearing.[1]

As the machine landed, hundreds of flat caps were tossed into the sky. People were shouting, 'Bully!' 'Astounding!' 'Magnificent!'

Grahame-White's machine never appeared. Later, I read that he had engine trouble after leaving Lichfield and was unable to finish the race.

'Father, let us go and see the Frenchman's machine,' I suggested.

There were several thousand people at the field that cool sunny morning in April, but we were able to wend our way through the crowd to Paulhan's Farman III.

The Frenchmen in the crowd were clamouring about Paulhan slapping him on the back and congratulating him in their language.

Glasses of champagne were being poured and lifted in his honour, in spite of the early hour.

In less than a year, the French had won two important prizes from the *Daily Mail* – Bleriot the previous summer for his Channel crossing and now Paulhan in this contest. 'Just look at those mad Frenchmen. They are drinking at this hour!' I said in astonishment.

'Do not begrudge them some pleasure, lad. Life is short and filled with difficulties and tragedies. Besides Paulhan will need the prize money as he is being sued by the Wright Brothers for supposedly stealing their aileron design,' Father replied.

Farman III Maximum speed: 37 mph
(From the Public Domain)

Thus, I cheered him for his superb accomplishment. Reflecting back, this was the first of many air races of any distance. The contest was much more than laps about pylons in a remote field. That day, the *Daily Mail* certainly achieved their intended goal of creating a public interest in flying machines.

We made our way through the crowd to the hundreds of bicycles along the side of the field. After some time, we found ours and began the journey home. I was so engrossed in the events of the day that I hardly noticed the beautiful countryside or the smell of spring in the air. About halfway to our destination, we stopped under some oak trees for a brief respite. Then, I informed Father, 'One day, I am going to fly!'

'My advice to you, lad, is to engage the company of other flyers as much as possible. If there are any menial tasks to be performed, be the first to do them. In this manner, you will be recognised and included in their endeavours.' He paused and added, 'Tomorrow has wings.'

Father's sage advice and prediction led me to some interesting and influential acquaintances. That day was a turning point for me and probably for Mr. Fletcher, as well. No doubt Mr. Fletcher was influenced by the air race, because he, Mr. Crossland, the Edwards brothers and A.V. Roe soon began to work on Mr. Fletcher's machine.

<><><>

IN JUNE, I HAD READ in the newspaper that Mr. Rolls had successfully flown a copy of a Wright *Flyer* he had built in England from Dover to Sangatte in France, and, after circling over French soil several times, he returned to Dover. He was given a hero's welcome upon landing for making the first non-stop round trip across the Channel.

However, July 1910 was the saddest month of my life prior to the war. On the morning of the 13th, I purchased a newspaper and was shocked to read the front-page headline and subsequent article. The article covered the death of Charles Rolls during a flying tournament at Bournemouth the previous day.

Charles Rolls
(From the Public Domain)

The machine had been fitted in France with a special tail section for additional control, which was the cause of the machine to fail in flight. Charles Rolls was the first Englishman to lose his life in an aerial accident. Unfortunately, he was just the first of several who would perish in a machine. Mr. Royce, in response to the tragic fate of his partner, changed the colour of one of the two ‘Rs’ in the Rolls-Royce insignia from red to black.[2]

This incident struck me hard, and it made me question the wisdom of my decision to become a flyer, especially in those early days of aviation when the machines were so fragile and difficult to control. My ambition was to be a flyer, however, so I had to put these thoughts aside.

There was a bright spot later that month. A.V. Roe had completed building Mr. Fletcher’s machine. It would be ready for aerial testing in August. The machine was based on the highly successful, and often copied, Farman III, designed and built by Henry and Maurice Farman. Charles Fletcher, Norman Crossland, the Edwards brothers and A.V. Roe all had an interest in the machine. Initially, it would have an Edwards motorcar engine, built by H.W. Crowley Company and modified for aerial usage by A.V. Roe. Later, it was to have a more powerful Empress engine, as A.V. Roe was interested in more powerful engines for his own machines and Mr. Fletcher had the resources to develop them.

Mr. Fletcher sent me to the Edwards brother’s garage in Bolton for a short time to help with the work. I was excited about the new design and looked forward to its first flight.

Chapter 4 – The Avro Farman

Manchester, England, 1910

IN AUGUST 1910, the Avro Farman, was ready for aerial testing. On the eighth of the month, it was brought to the Manchester Racecourse, which had sufficient open space for our purposes. Everyone could see my excitement over the new machine.

Edwards two-cylinder engine.
Designed by M.F. and W. Edwards.
Built by H.W. Crowley.
(From MOSI, Ref. A1997.36/6)

That afternoon, we prepared the machine for flight. Mr. Fletcher took the machine for its first practice. It seemed to lack power, and subsequently the propeller was changed. The change greatly improved its speed, but Mr. Fletcher underestimated the rate at which he was traveling. He attempted too sharp a turn, which caused the right wheels to buckle and the machine to suddenly collapse. He jumped from his seat and was unhurt, but it took us a couple of days to repair the broken framework, torn canvas and smashed stays of the machine.

After the repairs, another test proved the machine to be underpowered by the two-cylinder, two-stroke Edwards engine. With only 15 h.p., the machine never left the ground. This disappointed me, because it was another failed attempt by Mr. Fletcher to develop a successful machine. The Edwards engine was later sold as an Avro engine, through the Aviator's Storehouse catalogue, to Lillian Bland, the first woman aviator.

The Avro Farman with Joseph Tonge at the controls.
August 1910. (From MOSI, Ref. A1997.36/5)

In later trials, Mr. Fletcher used a seven-cylinder Empress rotary engine. The 50 h.p. engine lifted the machine off the ground. However, one flight landed in the River Irwell, which resulted in more repairs to the machine.

Everyone was confident of my abilities, to the extent that I was allowed to taxi the machine and eventually take it for a short flight. Taxiing was very tedious in those days. We turned the machine in the direction I was to move. One of the mechanics said, 'Be sure the oil tank is filled before you start the engine. These rotaries use almost as much oil as they do petrol.' He motioned for me to climb aboard the machine, and he did the same.

Another mechanic spun the propeller, and the rotary engine started on the first attempt. It quickly increased its speed to maximum power. The machine lurched forward, and its speed rapidly increased. The mechanic explained, 'There is limited rpm control on these engines, so to reduce the speed and prevent the machine from lifting off, throw the blip switch to turn off the engine. After you have slowed the machine sufficiently, throw the blip switch again to restart the engine. Be sure the engine does not stop rotating completely or the propeller will need to be spun again to restart the engine.'

He demonstrated the procedure several times with me, then he allowed me to take the controls. My first attempt was unsuccessful as the propeller stopped rotating sooner than I had anticipated. This required a mechanic to restart the engine. My subsequent attempts at taxiing were all successful. However, to move the machine any distance required frequent movement of the blip switch.

Empress seven-cylinder rotary engine.
(From MOSI, Ref. A1997.36/7)

In the short time we spent taxiing, the little Empress engine spewed out a considerable amount of oil and unburnt petrol, as we expected. This was in part due to the rotary design, where the cylinders rotated around a fixed crankshaft.[1] In this regard, I was thankful for the pusher design,[2] which positioned us away from the engine's slick rain. Thus, we avoided becoming oily pilots like all who flew machines with front-mounted rotaries. We still required protective attire from dirt and small stones, which often became aerial objects. Also, the machine's tail section required wiping after use as the engine always deposited considerable oil there.

I enjoyed being a pilot flying some short distances. I knew then that my decision to become a pilot was the right one.

There were several crashes at that time, but fortunately, no one was seriously injured. Despite the crashes, the Farman design was remarkable in many ways and before its time. It had wheeled landing gear instead of skids, and ailerons instead of wing warping (a system of pulleys and cables to twist the trailing edges of the flexible wings in opposite directions for lateral control).

One crash smashed the Racecourse manager's cucumber frames. It did not take very long before he approached Mr. Fletcher, and said, 'You will have to leave the Racecourse, as I can not have you endanger the people who work here.'

We wondered just how much of his concern was for the people and how much for his cucumbers. Nevertheless, Mr. Fletcher and I removed the machine to Brooklands, and he advertised the capabilities of the Empress engines.

Sadly, the Avro Farman was sold, without the engine, shortly after we delivered it to Brooklands. The sale made me think about my future at Empress

Chapter 5 – Changes

Manchester and Brooklands, England, 1910-1911

IN SEPTEMBER 1910, there was considerable discussion in *Flight* magazine between the Dreadnought Rotary Engine Syndicate and Mr. Fletcher. Dreadnought claimed Mr. Fletcher had used their design in his engines, and Mr. Fletcher countered that he had produced rotary engines for eight years. *Flight* challenged the Syndicate to produce the patent numbers of their design. They never did. As a result, there was no interruption in the use of Mr. Fletcher's engines, and Mr. Macfie, with assistance from Mr. Valentine as pilot, continued testing an Empress-powered machine at Brooklands through the rest of the year.

Macfie Empress Biplane
(Courtesy Michael H. Goodall)

I was disappointed about my situation; there was little for me to do regarding machines, now that Mr. Fletcher had sold his Avro Farman. But things were to change in 1911. Mr. Crossland, who had been in Mr. Fletcher's office, approached me at the works and said, 'Jack, I would like to speak to you about an opportunity that may interest you. Would you be willing to meet me at my company Friday evening?' he asked.

'Yes, Sir,' I replied, and we determined a suitable time to meet. I wondered what he had in mind, but he did not reveal his thoughts.

On Friday evening, Mr. Crossland met me at his office and motioned to me to sit in the chair across from his desk. He leaned forward and quickly moved to the reason for our meeting. 'Jack, as you know, I am deeply involved in motorcars and flying machines. I am in need of a good mechanic, like you, to help me. Would you be interested in working here at Crossland?'

I was pleasantly surprised, as I had not had any indication that he was in need of a mechanic. At the same time, I was a little dismayed in that he was not developing any flying machines that I knew about.

Mr. Crossland must have sensed my disappointment, because he quickly added, 'Jack, you are an excellent mechanic. For that reason, you would be the mechanic in charge of all the mechanics here at Crossland and Co. Automobile Engineers. Also, I would keep you connected with my associates in the Manchester Aero Club and help you become a pilot.'

This was quite a promotion for me, as I was 18 years of age at that time. We discussed my duties and wages, and I quickly accepted his offer. Finally, we set a date when I should begin working for him. We concluded, and I left his office.

Two weeks later I was at his works and began familiarising myself with his business. It became apparent that Mr. Crossland was connected to other motorcar interests. He frequently worked with Daimler, and the Calthorpe Motor Company, and he was involved with the Manchester Motor Club reliability trials. I helped him in this endeavour by taking several motorcars on cross-country reliability tests.

One day in 1911, Mr. Crossland called me into his office. 'Jack, Mr. Fletcher has an aero engine that needs repair. He no longer has a mechanic that can work on these engines, so he asked me to lend you to him for a while. When would you be available for him?'

'In three days, I am finishing some work and should be available on Wednesday, if that is satisfactory to you and Mr. Fletcher,' I answered.

At Empress, I supervised the work on the aero engine, and we soon had it running once again. To my surprise, Mr. Fletcher asked me to deliver it to its owner at Brooklands.

Aviation Sheds at Brooklands
(Courtesy of John Pulford)

Brooklands was a racing car circuit with its banked turns and a concrete roadway. The machines took off and landed on the field in the centre of the course. On my arrival, I noticed two machines on the far side, where there were groups of simple wooden sheds, which served as aviation schools and housed the machines, mechanics and flyers.

I asked them where Mr. Ducrocq was located. They pointed to a nearby shed, and I proceeded there.

I introduced myself to Mr. Ducrocq, and he asked me to install the engine in his machine. Some of the mechanics at Brooklands and I installed the engine and tested it. It ran very well, and Mr. Ducrocq was pleased. Then, in December 1911, with the work completed, I asked him if he needed a mechanic. I would be closer to machines at Brooklands than at Manchester, and I would have a greater possibility of obtaining my pilot's certificate. He took me on, and I became his chief mechanic. Thus, I began a new phase in my life, which would lead to my flying ticket.

Chapter 6 – Brooklands

Brooklands, England, 1912

MR. MAURICE DUCROCQ was an agent for Viale Engines and a demonstration pilot. He came to Brooklands in August 1910 to learn how to fly. In November, after 20 lessons, he received his certificate from the Royal Aero Club and was the first to graduate from the Hewlett-Blondeau Flight School.

Maurice Ducrocq
(From *Flight International*, 5 November 1913)

On at least one occasion, he said, 'As an aviator, I occupy a unique position. Accident of birth has made me a Frenchman; accident of business has kept me in England. The result of these two accidents and an intense desire to conquer the air has been that I am the only Frenchman who has taken his pilot's certificate in this country.'

Mr. Ducrocq acquired a Henry Farman machine and began his own flying school. He was of average height and size, and he looked quite the gentleman in his tweed suit with a dark waistcoat and matching tweed flat hat. He was the sales representative to Great Britain for aero engines produced by the Italian Spirito Mario

Viale. The Viale engine was a five-cylinder radial that produced 35 h.p., but it was also rated at 50 h.p.

In 1911, Ducrocq acquired a Viale engine, fitted to a Bleriot. Harold Barnwell had previously bought the engine from Ducrocq as part of an unsuccessful attempt to join A.V. Roe. When the arrangement fell through, Barnwell allowed Roe to continue using it. I assisted in fitting it to an Avro Type D machine.

At this time, an interesting machine that appeared at Brooklands was the Avro Type F, which was designed by A.V. Roe. Mr. Roe believed that open cockpits exposed pilots to collisions with birds and insects, so he built the Type F with an enclosed cabin.

The Type F's construction began in the Avro workshop in Manchester in January 1912. On completion in April, the machine was sent to Weybridge by train and was wheeled through the streets to Brooklands.

Avro Type F, 1912.
Maximum speed: about 65 mph.
(Courtesy of Michael H. Goodall)

The Viale 35 h.p. radial engine from the Type D machine was installed in the Type F machine. The installation went smoothly, but the subsequent engine test was another matter.

In those days, we used a very elementary method to test the engine's output. A rope was tied to the tail of the machine, and the other end of the rope tied to a spring balance of the type butchers used to weigh beef carcasses. Then the balance was attached to a second rope, with the other end of this rope attached to a stout post driven into the ground. When the engine was started, the balance indicated the forward thrust provided by the engine and propeller.

After lashing the machine to the ropes, I said to Mr. Hillstead, 'Climb into the cockpit and keep up the pressure in the petrol tank.' We had fitted the filler cap with a tyre valve, through which the tank pressure was maintained using an ordinary bicycle pump.

At that point, I had the engine started. As the engine increased in speed, it emitted a considerable amount of smoke, which went directly into the cockpit. The cockpit quickly filled with the smoke, and we could no longer see Mr. Hillstead. Poor Mr. Hillstead had to be dragged from the machine. We loosened his collar and put a wet rag on his forehead to revive him.

'I did not think being anaesthetized was part of my duties here,' Hillstead said, while rubbing his eyes after he regained consciousness.

After that incident, we devised more suitable methods for testing engines.

<><><>

ON MAY 3RD 1912, the machine was ready for aerial testing. Mr. Wilfred Parke made the first flight of the Type F, which also was the first flight of an enclosed monoplane. Flight-testing continued until May 17th, when a maximum altitude of 1,000 feet and a maximum speed of about 65 mph were reached.

Unfortunately, the Viale engine had a tendency to overheat after about 30 minutes. As a result, on May 25th, shortly after Mr. Parke took off from Brooklands to fly to Hendon, the engine failed and forced him to land at Weybridge. Parke hit a fence as he descended which caused the Avro F to turn over and crash. Damage was slight, but the machine still needed to be dismantled and returned to the Avro workshop at Brooklands for repair.

The Type F was designed for the British Army's 1912 Military Trials, but the army decided it wanted a two-seater. This made the single-seat Type F unacceptable, causing A.V. Roe to abandon the machine and to develop the Type G to meet the army's requirements.

Mr. Ducrocq also had the agency for Hanriot and arranged for the French machines to be built in Britain. He wished to enter two Hanriots in the 1912 Military Trials, and I, as his chief mechanic, made sure the machines were ready for flight-test.

Hanriot with 100 hp Gnome engine, 1912.
Maximum speed: 75 mph.
(From MOSI, Ref. A1997.36/5)

The pilots, Sippes and Bielovucic, approached me and asked, 'Are the machines ready?'

'Yes, they are,' I replied.

The initial flight-tests and the trials went well, but unfortunately, a Cody biplane, successfully passed all the tests and was declared winner of the competition. This surprised and disappointed me, as our Hanriots demonstrated better all-round performance.

I was very satisfied with my surroundings. Brooklands was one of the two centres of aviation in Britain before the Great War; the other was Hendon. After working for Mr. Ducrocq for some time, I approached him and asked him to teach me to be a pilot.

'Well, you are an excellent mechanic, and you are very likable. I would be glad to teach you to fly.'

He quickly came forth with a plan for me. The following weeks were certainly the most joyous events of 1912.

In the morning or evening, as the wind allowed, Mr. Ducrocq took me up in his Henry Farman machine. We both wore leather coats and helmets and flight goggles to withstand the elements. I had all I could do to remain calm and keep my excitement to myself. The engine ran smoothly. There was a sudden acceleration along the ground, then a steady lift. As we ascended, I became unaware of the speed. I felt a strange exhilaration and air rushing all around me. For my first lesson, I was taken to a considerable altitude and a significant distance from the aerodrome, which instilled no small amount of fear in me at first. This was quite different than the short hops I had taken in Mr. Fletcher's machine at Castle Irwell.

Mr. Ducrocq explained to me afterwards, 'It is necessary for me to take you to high altitudes so that you lose any fear and learn the possibilities of the machine's aerial capability.' [1]

After some days of this sort of flying, when my heartbeat was more normal, I was taken around the course at a lower altitude. In this manner, I was being prepared to understand the machine and its workings rather than my own feelings.

After three or four lessons of this type, Mr. Ducrocq said, 'Take the lever and feel with your hands how the machine responds.' It was his matter-of-fact manner, which put me at ease and marked Mr. Ducrocq as a born teacher and flyer.

After 11 lessons, I began to fly up and down the aerodrome, at first skimming the ground, then leaving it for short intervals, then finally I was in the air. In my twentieth lesson, Mr. Ducrocq had me perform the Royal Aero Club flying exam. Although there was no Club official to record my flight, I executed the certificate requirements: I flew without the aid of a passenger, made five figure eights at an average height of 200 feet, then landed within a circle of 50 yards diameter.[2] I do not hesitate to say that I could not have been taught better or more economically, in England or anywhere else, due to Mr. Ducrocq's excellent capabilities and his splendid machine.

At this juncture, well into autumn of 1912, I was ready to obtain a Royal Aero Club certificate. The certificate was my ticket to some very unusual experiences.

Chapter 7 - Early Air Races

Brooklands, England, 1912 and 1913

ON SUNDAY, NOVEMBER 14TH 1912, I made several circuits around the aerodrome, and I practised a few figure eights in the Maurice Farman owned by Ducrocq. The following evening, I again flew the figure eights circling the pylons, and I passed the first test to earn my *brevet* from the Royal Aero Club. I was ready to complete the remaining tests, but darkness overcame the field, forcing me to postpone my final flight until the morrow. On Tuesday, I completed my qualification flight by landing on the mark in the centre of the 50-yard circle chalked out on the grassy field. This landing, in no small way, surprised the spectators and officials observing my performance.

Mr. Ducrocq ran to where I had landed and pulled me from the cockpit before I could extract myself from the machine. Two of the mechanics were close behind him with a small table, glasses and a bottle of champagne. Once again, I experienced the mad French custom of imbibing at an early hour and at a less than favourable temperature. It seemed to impress them to no small measure that I had landed at the exact centre of the circle. '*Magnifique*,' they repeated several times, which surprised me as I had practised this landing until I had it well under control.

'Now practise those figure eights for the race on Saturday,' Mr. Ducrocq suggested.

However, I did not perform the figure of eights, as I would have liked. I could have held the pattern better. I needed more practise of these manoeuvres if I were to achieve any degree of success in the upcoming air races. On Thursday morning, I flew a few figure of eight circuits and made a short cross-country flight. The weather was good, however quite cold, and the sky was without a cloud. After noon, I flew a few more circuits in somewhat warmer conditions with a passenger on board and then another cross-country flight. I repeated these manoeuvres on Friday and finished with a high altitude flight over Weybridge. I was confident of the machine and my ability to utilise its capabilities. I was ready for my first air race on Saturday.

Mr. Ducrocq met me at my shed Friday afternoon and observed, 'The sky is quite dark, and sunset is several hours away. The weather may not support any races tomorrow.'

I was disappointed, because I was so anxious to test my skills in a race against other flyers. Yet, I knew from past experiences that Mr. Ducrocq probably was correct in his forecast.

Rain and high winds accompanied the sunrise on Saturday, which postponed the scheduled events. Late in the evening, the inclement weather left as quickly as it had appeared, and Sunday was a perfect day for the aerial contests. The sky was a bright blue and once again without a cloud, although the field had a considerable amount of mud.

I was scheduled for the first race and went to the field to hear the rules.

'The speed handicap requires two circuits around the pylons,' a short man in a striped shirt and flat hat announced. The distance between the pylons being less than a mile limited the speed of the machines due to the constant turning, hence the reason for the race being named the speed handicap.

Soon we were in the air on the first leg of the race. My machine responded well to the commands I gave it. It seemed like a thoroughbred ready to react to the slightest pressure on the bit. The gyroscopic forces from the rotating engine and propeller pulled me in the direction of the turns, and gave me no small advantage in the aerial competition.

At first, Tommy Sopwith pulled ahead of the rest of us in a machine of his own design and manufacture. Mr. Knight was along the side of me in his machine through most of the first lap. I had some difficulty in maintaining my intended course due to an inordinate amount of oil my goggles received from Tommy's engine.

On the second lap, having cleared the oil with my scarf, I pulled ahead of Tommy as we turned past the pylon. My machine shuddered as I banked to make a sharper turn than Tommy could. Thus, I achieved the advantage of being inside his circuit. I completed the race in 2 minutes 46 seconds, just 2 and 5/8 seconds faster than Tommy and 4 seconds faster than Mr. Knight. Messrs. Hawker, Spencer and Merriam, in their machines, followed us across the finish line. The Maurice Farman machine,[1] built by The Hanriot Company for Mr. Ducrocq, served me well, and I received a medal from the Manchester Aero Club for my achievement.

That evening, Mr. Ducrocq met me at my shed. 'Come with me to the Blue Bird to celebrate your victory,' he said.

We walked the short distance across the muddy field to the Blue Bird Cafe, where we joined other pilots and mechanics in celebration of the day's activities. After we sang several songs around the piano, they toasted my success. 'To the new pilot and his first win,' was about all I could hear over the laughter and song. At that point, we were too elated to fly, should we be called upon to do so.

It was times such as these at the Blue Bird that I developed lasting friendships and learnt the finer points of the many different machines at Brooklands.

The following Saturday, November 28th, there was a 'Bomb Dropping Competition' at the aerodrome. Eight other pilots and I entered this contest. I did not fare as well as I had the previous Sunday, but I enjoyed the day, nevertheless. Two days earlier, and three weeks after my twentieth birthday, I was presented with my pilot's certificate (number 368) from the Royal Aero Club. This certificate, and the aerial competitions that followed, put me in good stead for some grand opportunities at Brooklands and elsewhere.

In December 1912, and through early 1913, I had some failures and some successes in the competitions at Brooklands. At the turn of the year, stoppage of a petrol pipe prevented me from completing one competition. In January, however, I came in second in Mr. Ducrocq's Farman, and on March 24th 1913, I participated in the Easter Monday Aeroplane Handicap, again in Mr. Ducrocq's Farman, which was powered by a 50 h.p. Gnome engine. I won first place of six contestants in the race. However, 1913 was to be more eventful as I became more involved in machine development.

Chapter 8 – Messrs. Parsons and Coatalen

Brooklands, England, Late 1912 - 1913

Jack Alcock's business card.
(From the Public Domain)

TOWARDS THE END of 1912, after I had received my pilot's certificate, I decided to advertise myself as an aviator. I had a small card prepared to attract those interested in flying. As a result, Mr. J.G. Parsons had me test his Sommer machine. It was a Farman-influenced box kite type design. However, it had an all-steel frame. Many of them were built in Britain by Humber, under licence.

Mr. Parsons designed his own machine, which was different, as it was fitted with automatic stability devices. Initial tests found the machine somewhat deficient in power. For this reason, we decided to test it with a new Bristol propeller. Designing a stable machine was thought to be the solution to training pilots quickly, which would help expand the aviation industry.

The machine, with the new propeller, did not perform as well as Mr. Parsons had expected, so he had . Muller design a machine more suitable for testing his stability device. This machine also had the Parsons stabilizer. Jack Humphrys and I flew it in May.

Humphrys flew it without the Parsons Pendulum Paddle Wheel Stabilizer, and I flew it with the stabilizer. Surprisingly, the machine performed quite well, with its heavy car engine. It climbed to 1,500 feet in 20 minutes. The 40 h.p. Aster water-cooled engine was quite heavy, weighing 400 lbs. with its two radiators.

It was thought the machine and stabilizer would perform better with a lighter engine, so it was fitted in July with a 70-h.p. Gnome seven-cylinder rotary. In August, I flight-tested this machine. Its performance was better with the new engine, but the stabilizer did not meet Mr. Parsons' expectations. As a result, the machine was sold in September.

Sommer biplane.
Maximum speed: 60 mph
(From the Public Domain,)

Lawrence Sperry developed the first successful autopilot. Sperry designed a lightweight version of the marine gyroscopic system his father had invented and coupled it to the control surfaces of a Curtiss C-2 machine to maintain the flight direction. He demonstrated this apparatus in a hands-off flight at a Paris aerial show in June 1914.

Parsons' biplane with stabilizing device.
(Courtesy of Michael H. Goodall.)

Tommy Sopwith and Harry Hawker were familiar faces at Brooklands. We competed in several aerial contests there in December 1912 and in 1913. After these events, I frequently enjoyed their company at the Blue Bird, and we became close friends. Later, they would reappear in my life, again as competitors, but still good friends.

Tommy Sopwith was a handsome chap, a few years older than me. He had a normal frame, but he was very athletic. He was an expert ice skater and was on the national ice hockey team. He came from a well-to-do family, his father being a civil engineer. Tommy was educated as an engineer at Seafield Park engineering school. In December 1910, he won a £4,000 prize for the longest flight from England to the Continent in a British machine. He used the money to start the Sopwith School of Flying at Brooklands. From his success as an instructor, he and several other chaps established the Sopwith Aviation Company in June 1912 and began designing machines.

Harry Hawker also was a few years older than me. He was short and thin. He wore a flat cap, which for some reason he never shared, he frequently wore it with the peak pointed rearward. Harry was an Aussie mechanic who came to Brooklands and was ready to solo in just four days after his first flying lesson from Tommy Sopwith. Everyone at the aerodrome was astounded by Harry's natural abilities with machines. Tommy immediately offered him employment as an instructor in his flying school. Harry was so versatile that Tommy had him at sundry times employed as a mechanic, a manager, and eventually as a test pilot for the Sopwith machines Tommy designed. In October 1912, he won the British Michelin Cup for an endurance flight of 8 hours 23 minutes.

In March 1913, the three of us were at the Blue Bird, when Tommy asked, 'Have you read in the aviation journals about what is happening at Sunbeam?'

'Yes, I have,' answered Harry. 'The chief engineer, Louis Coatalen, whom they hired a few years ago to develop motorcar engines, is now developing an aero engine.'

'I am glad to see Sunbeam developing an engine. There are few British companies producing aero engines,' I pointed out.

'I find it heartening that they should do so. Since Charles Rolls died in an aerial accident, Henry Royce has been averse to developing aero engines. It is probable that Rolls-Royce would have made aero engines, and possibly aeroplanes by now if Rolls had lived,' Tommy added.

The three of us agreed that it was good to hear about Sunbeam, because it was another indicator that aviation was moving from small works to an industry in its own right. I enjoyed those days, testing machines for Mr. Ducrocq, instructing novice pilots, and competing in aerial contests.

<><><>

A MOST INTERESTING EVENT occurred on April 1st 1913. Lord Northcliffe, owner of the London *Daily Mail*, offered £10,000 to the first person to fly from the United States, Canada or Newfoundland to Great Britain or Ireland in less than 72 continuous hours. He did this to encourage flying and to interest the public in aviation.

Many readers of the *Daily Mail* thought this announcement was an April Fool's Day joke, whilst others thought it was a publicity stunt. The humorous magazine *Punch* responded by offering a prize for the first person to fly to Mars.

None of us thought a transatlantic flight was possible in 1913, but we did not discount the possibility of a successful flight across the Atlantic in the distant future. Certainly, none of us imagined that we would be competing against each other for the 'Atlantic Prize,' as Lord Northcliffe's challenge was later called. In 1914, a Martin Handyside monoplane was based at Brooklands, which was scheduled to be flown by Gustav Hamel in a transatlantic flight attempt in August but was cancelled due to Hamel's disappearance over the Channel in May and the outbreak of the Great War.[1]

A £10,000 prize was offered by the *Daily Mail* for completing a 1,000-mile circuit of Britain in one day and was won by M. Beaumont on July 20th 1911. From this prize, we knew Lord Northcliffe was serious about promoting aviation by offering such large sums for major aerial accomplishments.

<><><>

ANOTHER PILOT I KNEW at Brooklands was Freddie Raynham. Freddie was involved in flight-testing the Avro Type H hydro-aeroplane, later renamed the Type 503. It was the largest machine A.V. Roe had built up to that time, with a 50-foot wingspan. On May 28th 1913, we took off from the river near Shoreham Aerodrome in the Avro and quickly gained altitude in the 100 h.p. Gnome-powered machine. He cleared a nearby bridge by more than 100 feet. Two tests were successfully conducted that day, which demonstrated the machine's rapid climb capability, and its low landing speed.

The following day, Freddie had me go to Brighton with him as his mechanic. We arrived there at 2,000 feet and landed perfectly, after many spirals, outside Volke's shed, where the machine was anchored without assistance. After some additional testing, the machine was sold to the German Navy. Three more were built for the Royal Naval Air Service but were removed from their inventory by 1916.

Raynham, like Sopwith and Hawker, became another competitor for the Atlantic Prize years later.

<><><>

IN JULY, I READ where Sunbeam had reached a point in the development of their new 100 h.p. Crusader engine, as to how it would be flight-tested. According to the journals, it was to be tested in a water plane. As it turned out, on August 6th, the machine could not take off. This was due to a deficiency in the design of the machine and not attributed to the engine. The pontoons on the machine were of a conventional design, as used in floating bridges.

Radley-England Waterplane, the company that built the machine, apparently was unaware of the work of Glenn Curtiss and others in America. Curtiss realised that water produced a considerable amount of drag on a moving hull. In January 1913, he incorporated a stepped hull for his Model F. The hull of this water plane produced lift as it moved through the water. When it raised itself enough to expose the step, drag abruptly decreased, and the machine quickly rose more to glide on the small forward section with much less drag. The pilot could easily sense the transition. The quick rise and increase in speed told him that his machine was ready for take-off.

Sunbeam's chief engineer, Louis Coatalen, not willing to wait for machine modifications and experimentation, conceived a different test method. He appeared at Brooklands one day and approached me as I was working on a machine. He had a commanding presence, being in the prime of life, a sturdy frame and a square jaw. He was very direct, 'Jack, I am Louis Coatalen from Sunbeam Motor Car Company,' he began with a slight French accent. 'As you know, we at Sunbeam are entering the field of aero engines. We have had some delay due to the machine that we fitted our engine into for aerial testing. Hence, we have taken a different

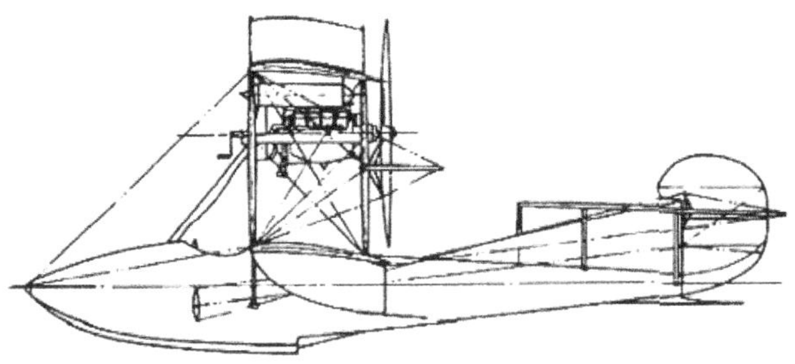

Curtiss Model F with stepped hull design
(From the Public Domain)

approach. I have decided to test our Crusader engine in a Maurice Farman aeroplane and would like you to be the test pilot.'

He continued, 'Jack, this is more than a test pilot's normal duties. As I mentioned previously, this is a new business venture for us, and we need to promote our engine. This will involve several entries in the more popular air races here in England. You certainly have demonstrated ability in that area. I am sure you would be willing to engage in such activities. My philosophy for air races is the same as for our motorcar races - racing improves the breed.'

Mr. Coatalen's offer took me by surprise. I immediately agreed to work for him, as it would be more than flight-testing a new engine.

The ground tests at Brooklands were somewhat basic in nature. It was the aerial testing, and the resultant design changes, that would ferret out the flaws to complete the development of the Crusader.

Chapter 9 – Test Pilot

Brooklands, England, Late 1913 - Spring 1914

THE CRUSADER ENGINE WAS READY for aerial testing. It seemed like a good engine, and I was anxious to see how well it performed aloft. Being water-cooled, it probably would not overheat, but there were other things that could malfunction in flight, as I quickly learnt.

I first flew it in Sunbeam's Farman in October 1913. There were some teething problems, and it was decided to send it to the Sunbeam works at Wolverhampton for repair. The engine was returned to Brooklands in December.

<><><>

IN THE YEARS 1912 - 1914, I spent many evenings at the Blue Bird Restaurant. It was there that I overheard Louis Bleriot, who now was producing monoplanes in France, say, 'My Bleriots are leading the way to the future. Monoplanes are more aerodynamically efficient and lighter than biplanes.' He was a striking figure, with his bushy eyebrows and handlebar mustache. Passionate about his machines, he frequently would hold his hand over his heart and proudly say, '*C'est magnifique!*'

I did not hear anyone counter Bleriot's claims and predictions, although many of us knew that the British government refused him a licence to sell his machines in Great Britain due to several structural failures of other monoplanes. Harry Hawker was with me that evening. Harry, being sometimes boisterous and temperamental, said, 'Louis, you need to be sure that your machines are safe! A machine that is aerodynamically efficient but crashes is no good.'

Blue Bird Restaurant, Brooklands
(Courtesy of John Pulford)

Bleriot said something to him in French, which I was sure was not very kind.

To avoid an argument, I steered Harry to another table in the outside café. Shortly after we were seated, Bob Dicker joined us. I had met him before at the Blue Bird. Bob was a thin, wiry chap, who was a mechanic and motorcycle racer. 'Jack, I saw you flying the Farman yesterday.'

'The flight went quite well,' I said.

'Maybe some day you would like to have a try at it on my motorcycle. It is quite a thrill, too,' he said with a little laugh.

'I would like that,' I said.

A week later, when I was at leisure, I visited Bob. He explained how his motorcycle operated and had me make small circles with it on the track near his shed. When he felt comfortable with my skills, he said, 'Take it onto the circuit.'

At first, I kept a fairly low speed to get the feel of the cycle on the concrete. I made one complete circuit under that assumption. Then I advanced the throttle to see how it would respond at higher speeds. It had sufficient power to exceed the speed of many flying machines, and I thoroughly enjoyed the thrill of riding it.

When I returned to the shed, Bob said, 'Jack, I clocked you at 50 to 60 miles per hour. Not many motorcyclists would go that fast the first time on the circuit. How did it feel?'

'Wonderful! It was an outstanding good time. I loved the sensation of traveling that fast.'

I frequently borrowed Bob's motorcycle to put in a fast lap or two. It was some going in those days as there were no shock absorbers or springs, which made the ride quite jolting at those speeds. Just before the Great War, I borrowed Bob's motorcycle to enter a hill climb contest. It was very competitive, and I managed to come in fourth.

Harold Potter
(From Collette Curry)

At Brooklands, I also met Harold Potter, who shared my interests in motorcycles, motorcar races, and flying.

I did not realise it then, but both Bob and Harold would be lifelong friends of mine. Bob was another person who would be involved with me in the Atlantic Prize. This day was just the beginning of many motorcycle and motorcar races I participated in while at Brooklands.

The Crusader engine, installed in the Farman, began a long period of intensive flight tests, which continued until the outbreak of the Great War the following summer. The engine performed flawlessly, and the flights became routine.

<><><>

ONE EVENING IN MARCH, after spending several hours at the Blue Bird, I walked across the aerodrome to where the machines usually landed. To my surprise at that late hour, I saw an old friend, F.L. Bailey, standing beside the field. He probably was enjoying the evening, as I was. 'It is a beautiful night for flying,' I remarked.

'Yes, it is. Would you consider taking me aloft this evening? I would like to see London from such a vantage point.'

'It shan't take long to ready the Farman, and we shall make a flight which will include more than just London,' I said.

Within the hour we were on an aerial adventure. I took the machine to 2,000 feet. After circling London for a brief period, I headed west as far as I thought the petrol could safely carry us, whilst enjoying the clear starry night. I then turned the machine to return to the aerodrome. The stars and moonlight were sufficient to judge the descent and a spiral landing. Upon landing and a short taxi run, my friend said, 'That was splendid.' Then he left me at the machine and went on his way.

A few days later I heard at the Blue Bird that he was surprised, by no small amount, when he learnt that he had been the first passenger in a cross-country flight at night. To the contrary, I was surprised that he and some others thought that being aloft at night was so dangerous and called my machine a 'scare plane.' I frequently flew on clear nights by following the roads below and the stars above.

<><><>

One day, my sister, Elsie, and a friend came to Brooklands. I asked, 'Would you like to go aloft in my Farman?'

'Yes, I would like to very much so,' Elsie replied.

However, her friend was hesitant, and finally she said, 'I will prepare a picnic lunch for the two of you, but I shall stay on the ground.'

'I can have the ma-chine ready by noontime,' I told Elsie.

Jack handing a picnic to his sister, Elsie (left), and another young lady.
(From Collette Curry)

As we increased our speed, during take-off, and as we gained altitude, Elsie became very quiet. Sensing her apprehension, I stayed at about 1,000 feet above the ground and pointed to some interesting things below. As she saw the houses and people, she seemed more relaxed and appeared to enjoy the flight. After about 20 minutes, I saw a familiar place to land and brought the machine to a stop in a grassy field. After enjoying our picnic, we returned to Brooklands.

Upon landing, Elsie remarked, 'This was exciting, indeed.'

No one at Brooklands mentioned to her anything about my machine being a scare plane, and Elsie returned to Manchester quite pleased with her experience at Brooklands.

Those were memorable days. In the spring and summer, I often spent time with friends at Brooklands. We would have picnics on the grounds and in the evenings spend time at the Blue Bird. During inclement weather, we would enjoy time together inside the converted shed and sing songs around the piano.

At that time, in the spring of 1914, Mr. Coatalen approached me with a new and very exciting proposal.

Jack (between two ladies on the right) and friends at Brooklands. (From Collette Curry)

Chapter 10 – Hendon

Hendon, and other Aerodromes, England, June – August 1914

ALTHOUGH BROOKLANDS WAS KNOWN as the centre for machine development, Hendon, also near London, was the centre for aerial sporting events. In June 1914, Hendon held an air race that attracted a crowd of 100,000 people, but more importantly to me, it showed that factors beyond a pilot's skill could determine the outcome of these events.

Mr. Coatalen approached me at my shed in Brooklands with a proposal. 'Jack, let me read an article about you from *The Aeroplane* magazine. 'Alcock's skill as an engineer, his diligence and perseverance and his soundness as a pilot found full compass.' We, at Sunbeam, also have recognised your abilities and would like you to enter the London-Manchester-London air race at Hendon using the aeroplane you have been testing. There is a significant purse for the winner, and the race will help promote our engine, especially as Mr. Winston Churchill, First Lord of the Admiralty, will be in attendance. The race is scheduled for the twentieth of this month.'

I was very excited about Mr. Coatalen's offer. Major events at Hendon attracted the latest machines and the best pilots, and it afforded me a chance to win a sizable prize. The *Daily Mail* offered a gold trophy, and Pratt offered £400 to the winner. The Anglo-American Oil Company also offered £250 to the winner and £100 to the pilot that placed second. Seventeen of us signed up for the race, including my friend Harry Hawker, with a newly designed 100 horsepower Sopwith. The journalists who covered these events, expected him to win.

The day before the race, eight pilots performed their speed trials at Brooklands and nine of us at Hendon with nine qualifying. However, the weather on the Saturday morning of the race was far from favourable. A thick mist rendered it impossible to see farther than about a half mile. Surprisingly, it was reported that there was good visibility at St. Albans and beyond, just 15 miles north.

Finally at 9:30 a.m., with the weather improving, the starter announced, 'Mr. Birchenough, please enter the contest.' Almost immediately, W. Birchenough took off in his 70-horsepower Maurice Farman with his mechanic as a passenger.

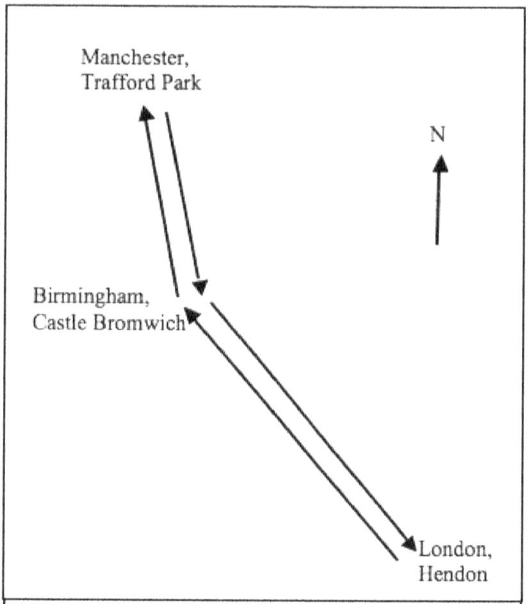

London - Manchester Air Race
(Sketch by J.F. Alcock)

Mr. Bjorklund was scheduled to take off next in his 50-horsepower Bleriot, but chose to disqualify himself because of 'indisposition.' I was next and took off in my Farman. About halfway across the aerodrome, my engine stopped suddenly. 'Oh blimey, not again,' I said aloud, reflecting back to previous mechanical problems I had experienced over the years. There was no immediate danger as there was sufficient field ahead for me to land, but this misfortune put me in a poor position to compete in the race.

Two mechanics in overalls came to where the Farman rolled to a stop and one said, 'Let us look at your engine.'

I agreed, and that turned out to be my best decision of the day.

After checking the engine, one mechanic said, 'It is a simple fix. There is a hole in the carburetor float. We shall have it repaired in short order. Grahame-White's works is just across the field.'

About two hours later, the repaired float was returned to the engine and the mechanic said, 'You are ready to have another go of it.'

I thanked them and thought it was still possible to compete in spite of this loss of time. The starter waved off L.A. Strange and Lord Carbery, then he motioned for me to take off. Harold Lane was my passenger, who would serve as my course finder. He was sturdy of mind and body and would be an outstanding asset in the race.

The flight to Birmingham was uneventful, except when Harold said, 'That is Birchenough heading back to Hendon.'

On my arrival at Birmingham, there were several people running to Lord Carbery's machine. He had landed about the same time as I had. A large, mustachioed mechanic came to me from Carbury's Bristol machine and said, 'Lord Carbery had an extremely hard landing and damaged his machine. He is not going on to Manchester, as it would take too long to make the necessary repairs. Neither is Louis Noel. He landed his Bleriot at the motorcar racecourse by mistake and damaged his machine, too. We just heard on the wireless that Birchenough returned to Hendon because his petrol tank was leaking very badly. He barely reached the aerodrome. I also heard Harry Hawker is out of the race, because he went way off course in the fog at Coventry. Brock and Carr have already landed here and are preparing to take off. Strange has already left.'

Of eight starters, only four of us remained just after the first leg! My flight to Manchester was uneventful. I landed at Trafford Park, near the city, and asked a small, wiry mechanic, who was refueling my plane, 'How is the race going?'

'Strange attempted to take off, but he damaged his propeller and chassis strut, and he was compelled to retire from the race. Brock and Carr already have left for Birmingham. You still have a chance, Jack. The contest is between just three of you.'

I thanked him and climbed into the cockpit whilst he spun the propeller. The take-off went well, and the flight was smooth to Birmingham. I said to Harold, 'We still can win.'

Then it happened! At about half the distance from Birmingham to Hendon, a severe storm moved into our path. The Farman shuddered severely from the strong winds, and I was worried that I would be forced to land, or worse yet, the machine would come apart.

When I thought all was lost, Harold exclaimed, 'There is a bright spot to the east.'

I thought, 'We need a bright spot, everything else is very dismal.' Then I headed towards it, hoping to circumvent the storm. As a result, there was some time lost, but we were able to finish the race.

After landing, a mechanic came to us and said, 'Brock is the winner in his Morane-Saulnier. Carr came in second, and then you chaps. No one else finished!'

We would have been close to Carr had it not been for the two hour delay at the start. I later transcribed the results into my journal for further consideration and analysis.

Journal: 20 June 1914				
Pilot	Machine	Engine, h.p., make	Flying Time, h.m.s.	Average Speed, m.p.h.
W.L. Brock	Morane-Saulnier monoplane	80 Gnome	4 42 26	69
R.H. Carr	Morane-Saulnier monoplane	80 Gnome	5 56 12	54
J. Alcock	Maurice Farman biplane	100 Sunbeam	7 56 17 *	45
L.A. Strange	Bleriot monoplane	80 Gnome	Damaged propeller and strut during takeoff at Manchester.	
H. Hawker	Sopwith biplane	100 Gnome	Returned to Hendon from Coventry after going off course in a fog.	
L. Noel	Morane-Saulnier monoplane	80 Gnome	Damaged machine from hard landing at Birmingham.	
J. Carbery	Bristol biplane	80 Le Rhone	Damaged machine from hard landing at Birmingham.	
W. Birchenough	Maurice Farman biplane	70 Renault	After 50 miles from Hendon returned due to petrol tank leak.	

* Includes the delay due to a faulty carbuerator.

The race was an object lesson on the reliability and strength of machines prior to the Great War. Of the eight contestants who started the race, four failed to finish because of mechanical difficulties or the frail nature of the machines. Interestingly, monoplanes were beginning to appear in these races more and more.

The carburetor problem made me think of an article I had read in *The Aeroplane* the year before. It read, ‘He has suddenly developed into a biplane pilot of the very first class. On a really up-to-date machine, Mr. Alcock would soon make a big name for himself.’ The article was very flattering. It made me think of other machines I might use in future races. The Morane-Saulnier monoplane, used by Brock and Carr, certainly was a fast machine.

Mr. Coatalen was pleased that I had completed the race, but he was upset with his engine after all the testing it had been through. He managed to say, ‘Jack, there will be other races, where we will do better.’

This was not to be, as I soon learnt.

<><><>

A WEEK LATER, on June 28th 1914, the newspapers announced the assassination of the Austrian Archduke Franz Ferdinand in Sarajevo. Due to the various alliances, the nations of Europe declared war with each other. When Germany invaded Belgium on August 3rd, that same day the British government sent an ultimatum to Germany demanding an immediate withdrawal. When the Germans refused, Britain declared war on Germany.

Meanwhile, on Tuesday, July 21st, Harold Lane and I left Brooklands to fly to Wolverhampton to deliver a magneto at the Sunbeam works. We ran into a thick fog, and I asked Harold to look for a place to land.

'There is an open field below off to the left,' Harold said.

I landed the machine in the field, which turned out to be at Snitterfield, near Warwick. When the fog cleared a little, I was able to get to Dunstall Racecourse at Wolverhampton and land there. We walked a short distance to the Sunbeam works.

At the works, several of the workmen greeted us. One said, 'Let us see your machine.' Several others joined him with the same request to see the Farman.

'We shall do one better. How would you like to see it fly?' I asked.

By now, several more workmen joined the first group, and they all answered with several cheers.

The following day, I had the Farman prepared for flight. For several days, I gave exhibitions of flying and passenger carrying to the Sunbeam workmen, who were keenly interested in the Maurice Farman machine, as it was fitted with a Sunbeam engine.

On Saturday, we started back for Brooklands in a severe wind. The machine shuddered from the gusts, and I saw that Harold had his hand over his mouth.

'Jack, I am airsick,' he said with a slight groan.

I landed at Snitterfield again to relieve Harold of his airsickness. At Snitterfield, we had a very hospitable reception, and, in return, I gave some flying exhibitions and returned to Brooklands on Monday.

<><><>

AFTER WAR BROKE OUT, the Brighton-Shoreham Aerodrome became a busy place. Additional hangars were being erected for pilots and machines, and there were numerous visits from the Admiralty and War Office officials to inspect the machines. Several machines were purchased there, including Mr. Coatalen's Sunbeam-powered Maurice Farman, which I had been flying. All the pilots offered their services to the Admiralty and War Office and were awaiting instructions. Being one of these pilots, I expected to join the Royal Naval Air Service and to be directed to the aerodrome at Eastchurch.

At Brooklands, all had changed, too. It was in the hands of the Royal Flying Corps (R.F.C.). One of the flyers I had known for some time came to me, shook his head and lamented, 'The Blue Bird is now the Canteen. Everyone has left. The Vickers, Bristol, and Sopwith sheds and many others have been turned over to the R.F.C.'

I noticed that at the back of the old Blue Bird, the Bleriot and the Martinsyde firms continued construction work. The other firms, I learnt, had set up works elsewhere as a result of orders from the government.

I remember sitting on a petrol tin, listening to 'It's a Long Way to Tipperary' on my gramophone. I was completely dejected. My beloved Sunbeam-powered Farman had been taken, and I was under orders to vacate my shed. My entire life for the last few years had been wrenched from me.

While awaiting further notification, I was hired by Pemberton Billing to test his P.B. IX. This gave me something to do to pass the time and forget the changes at Brooklands somewhat.

Pemberton Billing P.B. IX
(Courtesy Michael H. Goodall)

I made two short flights in the *Seven Day Bus*. The P.B. IX was called that, because it was constructed in seven days. I soon considered the machine unsafe, due to its hasty construction, and I refused to fly it again. This certainly was an inglorious end to my flying experiences at Brooklands. I was deeply saddened by my change of fortune.

Thus, a chapter in my life ended, and a new one began, which, unbeknown to me at the time, would take me closer to the Atlantic Prize.

Chapter 11 – Eastchurch

Eastchurch, England, November 1914 – December 1916

IN 1908, EASTCHURCH WAS NOTHING but a marsh on the Isle of Sheppey near London at the mouth of the Thames. The following year, Frank McClean acquired Stonepits Farm in the marshes and converted it to an aerodrome for the Aero Club of Great Britain, which later became the Royal Aero Club. After the outbreak of the Great War, Mr. McClean loaned his machines to the Royal Navy at Eastchurch to train officers in the skill of flying. I was one of the instructors, who arrived there in November 1914.

Training the necessary pilots for the Air Service was hard work and difficult at times owing to the serious shortages of machines. In late 1914 and early 1915, the chaos of producing pilots under these conditions was at its worst. As a Warrant-Officer Instructor at the Royal Flying School, I met Erroll Boyd in April 1915. 'Why would a Canadian want to join the Royal Naval Air Service?' I asked.

'I first became enamoured with flying a few years ago as a passenger in an aeroplane piloted by the American barnstormer, Lincoln Beachey. When the Great War erupted, I thought of serving as a pilot, but Canada had no training schools. So, I came to England to learn to fly. I attempted to enlist in the Royal Flying Corps, but I was turned down because of my colour blindness. I then applied to the R.N.A.S. and was accepted,' Boyd said.

'Splendid. Tomorrow we shall begin your lessons,' I replied.

The following morning, we met at the hangar. I asked the mechanics to provision the Wright biplane for flight. Soon we were aloft with me at the controls. 'If you can fly this machine, you can fly anything,' I explained.

'Why is this machine so difficult to fly?' he asked.

'It uses an early method of lateral flight control called wing warping, which is twisting of the wing tips. It is very sensitive and takes some time to get the feel of it. Therefore, you shall start on the ground and gradually work up to flying.' I brought the machine up to about 4,000 feet and circled the field to see how Boyd reacted to flight. He was smiling the entire time, which made me confident that he would be a good pupil.

I landed the machine and said, 'Take the controls and taxi the machine for a while.'

We taxied around the field some, before returning to the hangar. In this manner, he gained the skill for his first flight. He was quick to learn, and soon he was handling the machine splendidly. I moved him to flying a Short Brothers machine, which had a pusher propeller. He mastered that quickly and easily passed his flight test for his Royal Aero Club certificate.

<><><>

ONE EVENING when all of us had retired, we heard an explosion on the airfield. We ran outside our sheds to determine the cause of the noise and quickly saw the source of the problem. Above us, several German zeppelins were dropping bombs on the runway.

Boyd dashed to the hangar and soon was taxiing down the runway in a 50 h.p. Bristol biplane. During his take-off attempt, his landing gear dropped into a bomb crater causing his machine to turn over. Fortunately for Boyd, the four 20-pound bombs he carried did not explode. Having escaped injury, Boyd quickly moved to a new 80 h.p. Avro to continue his pursuit of the Germans. He was unable to climb above 4,500 feet and could only watch the Germans drop ballast and easily escape. In those days, our machines did not carry guns. We were expected to climb above the zeppelins and drop bombs on them, hoping to ignite and explode the hydrogen-filled machines.

Jack with Bristol Scout Type D.
Maximum speed: 94 mph.
(From Collette Curry)

Soon, Boyd was transferred across the Channel to No. 1 Wing at Dunkirk, France. In October 1915, his machine was hit by ground fire, causing him to glide to a crash landing in Holland. Holland, being a neutral country, allowed him to return to North America due to the injuries he suffered in the crash. He married a beautiful dancer, Evelyn Carberry, in New York City. Her friend, singer Al Jolson, hosted a sizable party for the newlyweds.

About this same time, I learnt that Mr. Fletcher had liquidated the Empress Motor Car and Aviation Company. This was sad news to me, as I had fond memories of working for Mr. Fletcher.

That evening, during dinner, the men complained about chasing zeppelins in the dark. ‘This is very dangerous; it is like being tossed into a lake at midnight,’ one pilot lamented.

'We appreciate Winston Churchill for creating the R.N.A.S., but we do not appreciate his promise that hostile zeppelins will be met by a swarm of hornets,' another pilot remarked.

Aviation writer C.G. Grey wrote, 'Churchill's directive to chase zeppelins in the dark is risky business. The Lord of the Admiralty needs to look at the number of wrecked aircraft from this endeavour.' The government later reported that in 89 flights of this sort, there were 20 aircraft partly or entirely wrecked. However, the practise continued.

Gradually, order was derived from the chaos, and we turned out some of the cleverest pilots on any front. Many, alas, are no more, but they rendered yeoman service at a critical time. Poor Warneford was one of my pupils, as was Colonel Mulock. Warneford was awarded the Victoria Cross for being the first to destroy a German zeppelin in an engagement near Ghent, Belgium. Sadly, a few days later, he and an American journalist died as a result of the right wings on his machine collapsing shortly after take-off.

Redford Mulock was credited with five aerial victories, which made him an ace. He received the Distinguished Service Order and was promoted to wing commander. Raymond Collinshaw, another pupil of mine, was also awarded the D.S.O. for his many victories. Like Erroll Boyd, he was one of Canada's most famous wartime pilots.

<><><>

IN AUGUST 1915, I had some respite from the work at Eastchurch. Brooklands had established the All Khaki motorcycle race for those in uniform during the Great War. On Saturday, the seventh, I entered eight of the 11 races held that day and won two of them. It was great enjoyment to be at Brooklands once again, but it was just a few days, as I had to return to duty shortly after the race.

Jack Alcock next to a 1914 Douglas
at the All Khaki Race
(From Collette Curry)

My commanding officer recommended me several times for a promotion. In November 1915, he noted my flying ability and surprised me that he also noted my loyalty to my country by my refusing from a private firm a salary of £800 a year as a test pilot, plus a £5 bonus for each machine I tested.

On December 27th 1915, I was admitted to Chatham Hospital under unusual circumstances; I had contracted measles. On December 29th 1915, I was made Flight Sub-Lieutenant and made head of the School of Aerobatics. I was excited to receive this honour, but disappointed that I was not given an assignment to fight the Germans, because I was considered more valuable as a flight instructor. Perhaps, I thought, if I were to catch pneumonia, I would receive an assignment to fight in the war. I remained an instructor at Eastchurch. My promotion, gave me some latitude to correct problems, as I saw fit.

Most of the new pilots were quite good and received their certificates quickly. My concern was that some of these chaps were quite cocky as well. I developed a technique that brought reality into their flying. One example was Dennis Wade. During his first flight on August 16th 1916, I said to him, 'Take the controls.' We were at about 1,000 feet, and I let him pilot the Voisen machine for about ten minutes. I could see he was overconfident, and I needed to apply my technique.

'Alright chap, I shall take over,' I said, and brought the machine up in altitude. As he looked off into the distance, I suddenly dropped the nose of the machine, dived towards the ground and performed a loop-the-loop.[1] I could not see his face as he was sitting in front of me, but when he turned towards me, I shall never forget the bulging eyes inside those goggles and his white knuckles from his firm grasp on the edge of the cockpit.

After we landed and were walking to the dining hall, Wade said, 'I thought my friends were pulling my leg, when they warned me that you perform some unnerving antics during your training sessions.'

'My antics are more than humour. They serve to illustrate that machines can be dangerous, if a pilot becomes overconfident at the controls,' I countered.

<><><>

THE PILOTS MET AT THE DINING HALL in the evenings. We enjoyed ale and bawdy songs around the piano. I must confess that I often participated and at times even led the festivities. However, we all longed for duty in the war zone. To relieve some of the tension, during those war years, the men would play pranks on each other. The most notorious of the lot were the mechanics, who were very innovative in this endeavour. Having put up with them for some time, I felt they needed a sound drubbing to bring them in line.

I endured a considerable amount of pranks from the mechanics. They let the air out of our tyres. They also produced a newspaper saying the Americans had won the Atlantic Prize for being the first to cross the Atlantic non-stop. It took me some time to determine this to be false. To respond in kind, I took one of the Curtiss machines to about 2,000 feet and circled the field. I continued to circle until I spotted my prey. Then, I shut down the engine, held up the nose until the propeller stopped and glided down onto the field and into the shed where the mechanics were working. The machine came to a stop exactly where I intended, a few feet from the startled men. I jumped from the machine and asked, 'Lads, why are you cowering on the floor in the back of the shed?' They never said a word. Later, when their antics became excessive once again, I repeated the performance.

Training pilots at Eastchurch was useful work, and very necessary also, but we chafed at its limitations. The desire to be out and in the war was always with us. We wanted a taste of what we were trained to do. Finally, in December 1916, I received a promotion to Flight Lieutenant, and I was assigned to the 2nd Wing of the R.N.A.S. near the Dardanelles when the Romanian front collapsed.

Chapter 12 – The Dardanelles

Isle of Lemnos, Aegean Sea, January – September 1917

EARLY IN THE GREAT WAR, Turkey signed a treaty with Germany and in the last days of October 1914, Turkish warships bombarded Russian ports on the Black Sea. As a result, France, Great Britain and Russia declared war on the Ottoman Empire (Turkey). In January 1915, Russia asked for an attack that would relieve them of Turkish pressure along the Black Sea.

Map of the Dardanelles Region.
(Sketch by J.F. Alcock)

During 1915, there were several unsuccessful attempts by the Allies to capture the Dardanelles, the strait that connects the Aegean Sea to the Sea of Marmora and the city of Constantinople.[1] The year 1916 was essentially a stalemate in this corner of the war. However, in January 1917, I was sent to Mudros, on the Isle of Lemnos, about 40 miles west of the western mouth of the Dardanelles. Mudros was the main base for Allied operations in the eastern Mediterranean. We were to attack the enemy in this region in order to contain their advance into Greece and elsewhere in Europe. Most of our activities were to fly anti-U-boat patrols and bombing raids in enemy territory.

<><><>

IN MARCH 1917, THE GERMANS moved twin-engine bombers and Halberstadt aeroplanes to the Macedonian Front to attack Allied supply depots in that region. As a result, R.F.C. and R.N.A.S. machines were gathered to counter this threat.

Sopwith Triplane. Maximum speed: 117 mph.
(From the Public Domain)

I had the only Sopwith Triplane at Mudros. It could out-fly any other machine in the region. I was ordered to fly to Stravos, near Salonika, and eventually join the other fighters. I attempted to land at a aerodrome there, but I overshot the field, landed in a ditch and wrecked the Triplane. As a result, I was left behind, whilst the others drove the Germans out of Macedonia.

<><><>

ON JUNE 8TH, WE WERE SURPRISED by the arrival of a large Handley Page 0/100 bomber at Mudros. This machine was to be used by the R.N.A.S. to bomb two German warships in the Bosphorus and to conduct a surprise bombing of Constantinople. On the night of July 9th, Commander Savory, with a crew of two, executed this great raid on the enemy.

Two days later, Savory was recalled to Britain to receive the D.S.O. for his daring raid. I took over the Handley Page and with Flight Lieutenants Wise and Gaskell, repeated the bombing of Constantinople and the Turkish bases at Adrianople, Panderma, and on the Dardanelles. On the raid of August 7th, we established a record by flying 600 miles in eight hours. These operations brought me my prised D.S.C.

No. 2 R.N.A.S. members at Mudros. Jack third from right. (From Collette Curry)

There was more than actual flying to be done. Machines were scarce. As a result, I called the other pilots together one morning, 'To increase the number of aircraft here at Mudros, we ought to build another scout with spare parts from damaged machines here at the base.'

Almost in unison, they answered my request, 'Let's go to it.'

I was very pleased with the enthusiastic response from the other pilots; thus we began the work the following day.

Two of the men conferred for a while, then one said, 'We have parts from several different types of Sopwith machines. We shall investigate how to put them together.'

After some discussion, we decided to take the forward fuselage and lower wings of a Sopwith Triplane, the upper wings of a Sopwith Pup and join them to a rear fuselage and tail section of original design. We fitted it with a 100 h.p. Gnome Monosoupape radial engine, but it was later replaced with a 110 h.p. Clerget 9Z radial engine. It was officially called the Alcock A.1 Scout, however, I usually called it the Sopwith Mouse.

Alcock Scout A.1, 1917.
(From Collette Curry)

I was not destined to fly the Mouse, as I was taken prisoner by the Turks whilst flying a bombing raid before the machine was completed. In the civil jail at Constantinople, on October 18th 1917, I received a coded message from Wing

Captain Scarlet, which read: 'Your baby was taken for an airing, but is still having trouble with teeth. She has now been fitted with new clothing. Now, a great improvement in health.'

After the war, I saw Flight Lieutenant Starbuck's logbook which records a flight in the Mouse on March 25th 1918, which read: 'First flight on the A1, (48 minutes, 5,500 ft. Mudros area), tried stunting it. Cartwheeled both ways and spun to the right. She came out of everything alright. This type of machine has never been stunted before. Very nice to fly. A bit nose heavy and heavy on the controls, but quite quick on answering.'

On April 2nd 1918, he made a second flight, of 20 minutes, 5,000 ft., in the Mudros area. He records, 'Practice flight and test gun and gun sight. Gun jammed 3rd position. Sight OK. Pressure gave out, had to pump to get back, tap at bottom of pump had slackened off.'

His next entry, on '2 April, 5.15 - 7.00 p.m., 1 hour 50 minutes, 11,000 feet. Mudros, Chanack down the Dardanelles to Cape Heller then home to Mudros. I set out with Wiser on a Camel to pick up 2 DH.4s over Chanack. When we got there light was failing so we returned and did not see the DH.4s. Wiser landed first. I followed, and just as I stopped, one DH was seen coming in from the opposite direction and down wind. I jumped out, and my machine was hastily pulled out of the way, but we were not quick enough, for the DH, which coming up the aerodrome at about 70 knots, crashed straight through my bus and knocked it to pieces and smashed itself up pretty badly. Luckily no one was hurt.'

Handley Page 0/100
(From the Public Domain)

In September 1917, my bad day arrived. The morning of September 30th, all was quiet at the base until we left the dining hall when suddenly three planes appeared high overhead.

'Look, those are German machines!' one of the pilots shouted.

'Let us go to it!' I said and we scrambled for our machines, still in our pyjamas, hoping to engage them before they were out of sight. I quickly brought a Sopwith Camel in pursuit of the interlopers. I caught up with one. At first, we circled to the right.[2] Each of us tried to get behind and above the other. Soon I discovered I was not meeting a beginner, as he had not the slightest intention of breaking off the fight. He was traveling in a machine, which turned beautifully. However, Tommy Sopwith's new model, the Camel, was better at rising, and I succeeded at last in getting above and beyond my German waltzing partner. The chap was full of cheek, and when we were down to about 3,000 feet, he merrily waved to me.

The highly successful Sopwith Camel
Approximately 5500 were produced in the Great War. Maximum speed: 115 mph.
(From the Public Domain)

The circles, which we made around one another, were so narrow that their diameter was probably no more than 300 feet. I had time to take a good look at him.

When he had come down to about 300 feet, he tried to escape by flying a zigzag course, which is a difficult target. That was my most favourable moment. I followed him from an altitude of 250 feet to 150 feet, firing all the time. My synchronised machine guns worked flawlessly.[3] By this time, my companions had joined the fight, and his fate was inevitable. Having been shot several times, he failed to maintain altitude and crashed into the sea below.[4]

For this action, I was later awarded the Distinguished Service Cross, which read: 'Flt. Lieut. John William Alcock, R.N.A.S. For the great skill, judgment and dash displayed by him off Mudros on September 30th 1917, in a successful attack on three enemy seaplanes, two of which were brought down in the sea.'

I should have retired after the encounter with the Germans, but, instead, three of us took the Handley Page for a bombing mission that night. While flying over enemy lines, the reduction gear on one of the propellers sheared off. We tried to keep going on one engine and managed to cover 60 miles in that condition.

We were forced to descend into the sea off Suvla Bay, on the Gallipoli Peninsula. Thus, my flying in the Great War ended.

Chapter 13 - Prisoner of War

Turkey, September 1917- November 1918

WE HUNG ON TO THE WRECK of our machine as best we might, but the Turks quickly located us and began sniping. As Johnny Turk began to get the range, Lieutenants H.R. Aird and S.J. Wise and I struck out for the land, swimming cautiously towards the rocks, amongst which we eventually hid from our pursuers.

All night we lay concealed, until 10 o'clock on the following morning. How long we could have eluded the enemy I would not like to say, but we were utterly worn out. I said to Wise, 'You have signs of nervous exhaustion. We need to consider handing ourselves over to the Turks, rather than have you get worse.'

Before he could respond, Aird said, 'We should surrender. Discretion is the better part of valour.'

Wise agreed. Upon seeing three burly Turks, we stepped from our hiding place and called to them. They pounced on us with little ceremony, stripping us of everything as a preliminary precaution. They took everything, and in somewhat embarrassing circumstances without our clothes, we were ruthlessly prodded towards the Turkish camp.

Fatigued and distressed in mind and body, we were much relieved to receive a courteous reception from the Turkish commandant. Through his interpreter, he said, 'Gentlemen, welcome to our camp. You shall be treated well here.'

I was doubtful of his words, until he promptly punished our brutal captors by revoking their leave and returning to us all our belongings. Save a few pounds, all was handed back - clothes and the contents of our pockets down to the smallest item. We were well fed and kindly treated and quickly recovered from the trying circumstances of our adventure.

We were sent to Constantinople, where Djavid Bey,[1] the Turkish Finance Minister, received us. He promised us hospitable treatment, so, well content, we went for temporary detention to a local jail.

We were to be moved on the morrow. When this did not occur, I asked the jailer, 'When are we to be moved?'

'Tomorrow,' he answered.

Day after day, 'Tomorrow,' was always his answer. True to Turkish tradition tomorrow was always on the horizon.

'How much longer will this place be our home?' my two companions asked.

'I cannot say. You now know the Turkish way,' I answered.

During our stay at Constantinople, I wrote the following letter to my uncle describing what had happened to us:

Prisoners of War,
CONSTANTINOPLE,
TURKEY.
15th October, 1917.

My Dear Uncle Alfred,

Well by now you will have heard we are prisoners of war and are very fit and healthy.

I had the rottenest luck in the world, my propeller and reduction gear burst about 1½ hours after I had started on a raid, I tried to get home but unable to, I landed safely in the sea after some effort, sat on the machine for over two hours making signals; why I wasn't picked up is a mystery to me. At last I had to leave the bus, which was sinking, to get ashore; was in the water well over an hour.

After all this I spent the night on the rocks shivering, and was made prisoners 10 o'clock the next day.

Since we have been prisoners, we have been treated very well indeed. Time passes very slowly now; would give anything for some work.

Will you arrange for all my gear to be sent home and my camera & photos, also photos of my bus.

Well Uncle, I shall want plenty of cheering up now so write often giving me the news of the new list of promotions and honours.

You might let me know who tested the bus & if she was any good.

I would give anything to be at Mudros, being prisoner of war does not agree with me, although I am treated very kindly & get plenty of food.

When you see the Capt., Give him my apologies. Give all my love to the boys and wishing them plenty of luck.

Will now close with fondest love,

Your affectionate Nephew,
Jack.

Have all our gear sent to our home addresses.

Finally, after a month at Constantinople, a physician and an interpreter came by our cell to examine us, and our fellow prisoners. The following day, the jailer came by our cell, unlocked the door and brought us to his supervisor, another burly Turk, with the same interpreter.

'Later today you and your companions will be removed to another camp.' That was all he had to say.

Indeed, we were sent to a happier place, a camp at Kedos, near the Hermus River,[2] arriving there on November 3rd. We had a good time, with freedom and abundant food. Lieutenants Aird and Wise and I were housed with some hundred prisoners of war, mostly from Scottish regiments.

Our only concern was how to pass the time. The three of us longed to be back in the aerial business again but realised that it was impossible in our current circumstances. I knew that it was important to our health and spirits to keep a stiff upper lip during our imprisonment. The commandant, Adhem Bey, was a really civilised type and possessed a sense of humour, unfortunately rare in a Turk. Doctor Spackman could speak a certain amount of Turkish and was willing to act as an interpreter. Therefore, with Spackman, I approached Adhem Bey with a suggestion, 'Is it alright with you if we help with the construction?'

'What construction?' the thin little man asked, with one eyebrow raised, not understanding that I meant the building already under construction.

'We would like to help building the bandstand in front of the barracks. The work will help us to deal with the boredom,' I replied.

'You may do these things,' he simply said.

I did everything I could for the men. I helped build the structures, and I supervised and catered the daily mess of 24 officers. Cooking had always been my first hobby, and at the camp, I made meringues by beating the whites of eggs. The men took up some sports. I played full back on the hockey team and joined every baseball game, a sport introduced by the Canadians. Occasionally, I had to refrain from these activities, due to bouts with malaria, which Spackman helped me through.

We were allowed to write letters, although this was not something I did often. I wrote Fred Mosely, my fellow Mancunian, 'Here I am doing time all through a burst propeller.' Despite the freedom we were allotted, I resented the wasteful days and longed for the days when once again I could be aloft in the latest machines. These times, however, granted me the opportunity to think about the future. I thought of the machines we were flying and the great distances they were capable of traveling. I became convinced that it was possible, with one of them, to fly across the Atlantic non-stop by some northern route. After one of these musings, I wrote home, 'Tell Coatalen I shall be ready for any big stunt after the war.' About that time, I learnt I had been promoted to Flight Commander in December 1917. (Later, this rank became Captain in the Royal Air Force.)

In July 1918, we began building a stage for a theatre on the edge of the orchard. It had a sloped floor, back-scene frames, wings and a proscenium.[3] The theatre provided the most amusement for us. The men put on 'Twelfth Night,' 'The Maid of the Mountain' and 'Chu Chin Chow.' There was no end of talent among the men, and the Turkish officials always came to these entertainments. One of which, I shall always remember as long as I live. It was the evening of September 27th, and the bill was 'Theodore,' a new piece prepared by our repertory company. The theatre was

Drawing of Kedos Barracks
(From Sir Leonard Wooley, *From Kastamuni to Kedos*, 1921)

crowded, and the Turks were in their accustomed place, when halfway through the piece I heard a cry of, 'Fire!'

It was not the theatre, but the town itself that was alight, and the subsequent conflagration beggared description. As I say, none will forget the sight. The fire started in a house next to the *khan*,[4] by a woman making peckmez,[5] and spread rapidly. The *khan* itself was one of the first buildings to ignite, and one of our men, who was there alone, had a narrow escape.

As we reached the scene, a British orderly told us, 'The fire has spread across the road, has destroyed one house, and it is advancing on more houses.' He then ran off to join some other orderlies. They were bringing a local fire engine with hand pumps and a leaking hose into place, but it had little effect.

Once it was determined that no lives were in danger, we began to salvage property from the houses occupied by our men. Those not engaged in this activity began demolition work on the structures to stop the fire from spreading further, including demolishing the mosque, to save the surrounding houses.

The commandant came to the scene and said to his countrymen, 'Help the British officers.' When he left the scene, the Turkish men did nothing and soon left, leaving the salvage work to the women. As a result, they cursed their men folk and worked in a perfectly random manner. Valuables were left to burn while they toiled away to remove pots and baskets of apples.

After the mosque was demolished, our crew went to help the women. Unfortunately, not all of the Turks understood our intentions. Some of the British officers were stoned by the local countrymen, until they realised, we were salvaging their stuff as well as ours. All night long, we were carrying off bedding, copper pots and sacks of grain and flour, to be claimed later by their owners.

Most of town was burned to the ground and blazed throughout the night without let or hindrance. Where a western people would have hurried to stem the flames, these Turks fled from the scene, wringing their hands in abject terror. As a result, the town was destroyed, and there was no more camp and no more theatre for us.

The following morning, Lieutenant Aird gave me the fire's toll, 'Jack, 2,000 houses were destroyed by the fire. There are about 300 left.'

After the fire, we lived in the open for a month.

We were to be moved up country to another and larger camp, but nothing was done. I went to the camp commander, 'When are we to be moved?' I asked.

Again, I received the typical Turk response, 'Tomorrow.'

At last, instructions were given by our captors, 'Bulgaria has surrendered.'

Master Turk saw the red lamp of defeat so clearly that we were ignored. When the inevitable end did come, we left Turkey on November 18th, after 13 months captivity, which was on the whole a very good time.

There was, however, tragedy enough in Turkey, but it was rather for the men than for the officers. The Turk, who cares little for his own men, cared nothing for ours, and from sickness and neglect, hunger and brutality, three quarters of them died miserably. However, there was a lot of mutual respect between aircrew officers on both sides. To that end, we were sometimes invited to join Turkish officers for a meal and drink at the Officers Mess. Thus, knowing what the men endured, we could scarcely complain about our inconveniences.

With my imprisonment behind me, I could proceed with my plan.

Chapter 14 – Meeting at Vickers

Manchester and Weybridge, England, December 1918 – May 1919

AFTER I WAS RELEASED from prison, I travelled to Smyrna, Alexandria, and Cairo, arriving at England on December 16th in time for Christmas, which some of us never expected to see again in our native land.

Christmas was a joyous time. Many inexpensive and hand-made gifts were exchanged. Socks were knitted, pies were made and given as presents, and cards were delivered. The main gift many gave to one another was the gift of time, a walk in the village park, or in the countryside. It was at this time that I took a walk with my father. Father told me about Mum, my sister and brothers. He spoke some about his job at the newspaper and about being a veterinary surgeon, which he learnt by serving time with an experienced vet. Then, he said, 'Jack, it is splendid to have you home. It has been quite a while since we were together.'

'Yes. It has been some time,' I replied.

'Your brother, Edward, also has followed your life quite closely and is planning to work in the aviation field.[1] What are your thoughts on this and what do you intend to do, now that you are back in England?'

'At present, there are many pilots and mechanics looking for work, now that the war is over, but eventually we shall see aviation take its place alongside other industries. I need to complete my duties with the Royal Air Force and pursue what I have been thinking about for some time, the Atlantic Prize,' I said.

'If anyone can fly the Atlantic non-stop, it is you. Not many believe that it can be done, but with the machines developed in the war, with increased range and improved engine reliability, I believe it is possible,' Father said.

We left the park and walked home in silence. My mind was on the future, and how I would pursue my dream.

As soon as it was possible, I met with General Scarlett, and I was demobilized on March 10th 1919. At last, I was free to attack the big problem of crossing the Atlantic. When I arrived at Brooklands, manufacturers Vickers, Sopwith, and Martinsyde were still there. Sopwith and Martinsyde had already declared their interest in the Atlantic Prize, which Lord Northcliff resurrected after the war. Also, Handley Page had entered the competition. Handley Page was well known for flying their machine in stages from England to Turkey, a 2,000-mile trip. This flight, which I had welcomed at Lemnos, clearly demonstrated a long-distance flight was feasible.

With three manufacturers already entering the competition, I decided to approach Vickers. I was confident that they would consider my proposal as I had been at Brooklands as an apprentice before the war, had been a flying instructor at Eastchurch at the beginning of the Great War, served on the Eastern Front, and had 4,500 flying hours to my credit. All of this, I included in a letter to Vickers, and a meeting was scheduled.

To my surprise and delight, when I walked into a conference room at Vickers, I was greeted by two of my acquaintances from Brooklands, Archie Knight and Percy Maxwell Muller. I refrained from initiating the conversation, until I could determine their reaction to my proposal.

'Jack, we understand you are interested in flying the Atlantic. We are deeply interested as well and are willing to support you in this venture,' Maxwell, the aviation superintendent at Vickers said.

'We would like you to have a go of it,' added Knight, the company's works manager.

'I am on it any old time!' I quickly replied.

I was pleasantly surprised by what followed. We shook hands, and the meeting ended. There was no ceremony, no discussion of schedule and no cost estimates. It was clearly understood that we were now in a major competition with at least three other companies. There was no time to waste.

I met with chief designer Reginald Kirshaw Pierson, and he revealed his thoughts on the machine best suited for our adventure. 'As you know, Jack, Vickers was contracted about a year before the war ended to develop a long-range heavy bomber. We called it the Vimy, after the Battle of Vimy Ridge, which occurred in France during the war. In an all-out effort, we built the first of three prototypes in less than four months. We had engine supply difficulties, tried several different engines, but finally settled on Rolls-Royce Eagle VIII engines. Only three machines were delivered before the war ended, and none were ever used for their intended purpose of night bombing Germany.'

'I have studied the charts and the shortest route across the Atlantic is from Newfoundland to Ireland, a distance of almost 2,000 miles. Is the Vimy capable of this range? I thought its capability was much less,' I asked.

'Its range is only 900 miles, but it can be modified for much more. I shall have extra petrol tanks installed and dispensed with the bomb racks. I believe it will have a range of 2,500 miles,' Rex Pierson answered.

We were still short another key component – a good navigator. It was entirely by chance that I met Arthur Whitten Brown one morning in March. I walked into Percy Muller's office and saw two men having tea with Mr. Muller. He introduced them to me. They were both R.A.F. officers, who were there to discuss radiators with Mr. Muller, our superintendent. With that short introduction, Mr. Muller was called out of the office, leaving me alone with Arthur Brown and his companion. Brown was of medium height rather slightly built with fine blue eyes, a cultured intellectual face, and wavy brown hair with some grey.

Brown and I talked for a few minutes about the war. Then he asked me, 'What is your role here at Weybridge?'

'I am involved with Messrs. Vickers in the Atlantic competition. I am to be the pilot of the machine now being assembled for that venture. The company does not want any publicity about it until we attempt the flight,' I said.

At that point in our conversation, Brown sat up straight and became more engaged.

'As a prisoner of war, I first took up seriously the study of aerial navigation. I decided to learn whatever might be learnt about that science. Aerial navigation for long distance flights over the ocean is considerably more difficult than nautical navigation. A serious drift caused by the wind, may take an aeroplane 40 or more miles off the course in each hour's flying, according to the direction and strength of the wind. In cloudy weather, or at night, a change in the wind can alter the drift without the knowledge of the navigator. Hence, one must observe the drift at all possible times,' Brown said.

'So, what would you propose for a transatlantic flight?' I asked.

'There are three methods of aerial navigation. I would use a form of dead reckoning and astronomical observation. Neither should be used alone. The third, directional wireless telegraphy, is not reliable, so I would not recommend it.' Brown answered.

'Dead reckoning is the simplest method of navigation, and under favourable conditions, it gives a high degree of accuracy. The air speed metre reading,[2] corrected for altitude,[3] provides the True Air Speed,' he explained. 'The compass reading provides the direction. If a magnetic compass, rather than a gyrocompass, is used, the direction must be corrected from magnetic to true north, which is shown on the navigation charts.'

'The True Air Speed and true north only provide the speed and direction of the machine relative to the air. If the air is moving, that is, if there is a wind, the speed and probably the direction of the machine, relative to the sea below, are different.'

'So, how do you account for the effects of the wind? It is a much different problem than flying over land.' I asked.

'To determine the speed, and the direction, relative to the sea, an instrument known as a Drift Indicator, or Drift-Bearing Plate, is used. One looks through the centre of the indicator at whitecaps, icebergs, ships, or other slow-moving objects visible on the surface of the ocean and turns the face of the dial until these objects are moving parallel to a wire on the dial. The angle which the wire makes with the direction in which the nose of the aeroplane is heading is the angle of drift,' Brown answered.

'And how do you determine the surface speed?' I asked.

'The surface speed can be measured by observing the time taken in passing over any fixed or very slowly moving object, while a certain angular distance is described – this being found by suitable sights, attached to the Drift Indicator, From the result, considered in conjunction with the height of the aeroplane, the actual speed over the surface is calculated. This speed will be in the direction shown by the wire of the Drift Indicator, [4] he answered.

To clarify his words, he drew a sketch of the different speeds and directions:

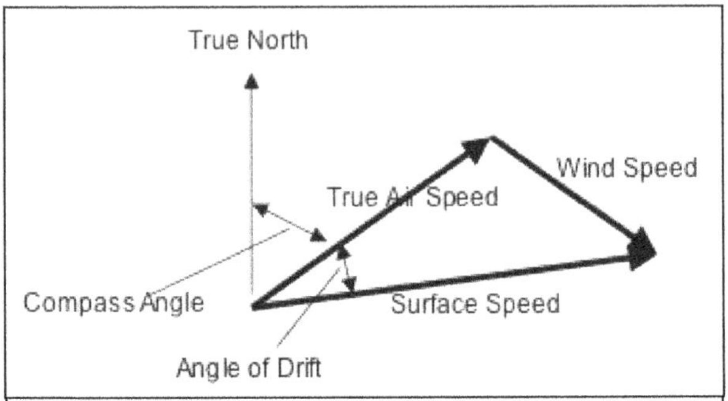

Flight Navigation Diagram
(Drawn by J.F. Alcock)

He continued, 'Astronomical observation is more involved and requires a 'fix' on the sun, moon or stars using a sextant to measure, simultaneously, the angular height of at least two of these objects above the horizon. If the horizon is not visible, a spirit level can be used instead, but it is less accurate. Two lines are drawn on the charts, one for each reading, and where they intersect is the position. The lines are determined from…'

He soon got beyond me. I concluded that he knew a good deal about it. I then asked him, 'Would you care to fly the Atlantic with me?'

'I would be delighted to,' he answered.

'I would like to show you something,' I said. I took him out to the works and showed him the Vimy on its assembly trestles. It was quite impressive with its 68-foot wingspan and the maze of spruce spars, steel tubing and sundry other parts.

'It looks like you are adding a significant amount of petrol tanks,' he observed.

'Yes, several more are needed for the flight. We need to carry 865 gallons of petrol. It is the thirteenth Vimy off the production line, my lucky number,' I noted.

'Will the Vimy be able to take off with the extra weight?'

'Messrs. Vickers and Rex Pierson believe that it will. It shall be difficult; it will weigh 13,000 pounds at take-off,' I answered. 'Notice the cockpit has been modified with a four-foot-wide bench so we can sit side by side for better communication,' I added.

Brown had nothing more to say. We left the works and agreed to meet to plan our labour to ready the Vimy for its long journey.

A short time after I met Brown, whilst I was at the works to inspect the Vimy, I learnt that my old friend Bob Dicker from Brooklands was working on the Vimy. On that particular day, I was dressed in my naval flyer's finery, as I had just met with some dignitaries regarding the transatlantic attempt. As I approached the machine, I saw Dicker had his head down in the cockpit. Instead of warning him of my presence, I simply said, 'Bob, I want you to go to Newfoundland with the Vimy.'

Before removing his head from the cockpit, he quickly said, 'Yes.' He was quite surprised to see me after all these years, especially in my uniform.

I then explained to him that I had been engaged by Vickers to make the transatlantic attempt.

I did not have time for meetings, as there was much organisation required to prepare for the flight. Therefore, I had Brown deal with the Air Ministry, which he did very effectively. Negotiation with the Air Ministry was difficult, because they believed that the future lay with airships and not with machines. Nevertheless, Brown was able to get them to agree to lend us a compass, a wireless transmitter and a receiver.

Vickers Vimy Mk IV with two Rolls-Royce Eagle VIII engines. (From the Public Domain)

After Brown informed me of his progress with the instruments, I asked him, 'Have you procured the petrol and oil needed?'

'I am working on it. There are several companies I need to contact. Also, we need to be fitted for our flying suits. I have scheduled us to be fitted at Burberry's,' he answered.

I was pleased that Burberry's was making the flying suits, as they had a fine reputation having been the supplier of clothing for Roald Amundsen and Robert Scott for their Antarctic expeditions. It took several trips, however, to get the electrically heated flying suits fitted.

The most difficult task, for Brown at least, was to postpone his marriage to Kathleen Kennedy. Her father, Major Kennedy, seeing the transatlantic flight as a patriotic venture, supported us whenever possible. Undoubtedly, he helped

Lucky Jim
(From MOSI, Ref. 1991.437.5)

Brown placate his daughter. Brown took his fiancée to the odd dance[5] and to a West End show[6], and she reciprocated with a small stuffed black cat, 'Twinkletoes.' I was not to be outdone, and produced another mascot, 'Lucky Jim.' We brought both 'cats' with us on our transatlantic flight.

W.J. Richards, a wireless officer, hired by Messrs. Vickers to assist Brown by installing the wireless, came to me to discuss the matter. 'We are mounting a small propeller under the right wing to drive a generator to power the wireless.'

'I don't want any bloody wireless. All I want is a pair of wings and a couple of engines,' I said. I was frustrated by all the space the wireless would take, to say nothing about the additional weight. Wireless sets were not very reliable, which made me question its value.

Brown finally convinced me, 'We may need it to send an S.O.S. to a ship along the route, if we have a problem.'

On Good Friday, we tried the Vimy thoroughly and found it so satisfactory that it was disassembled and packed for transatlantic shipment without delay. Strikes were the order of the day, however, and we were hung up for a considerable time at the docks. Besides our two selves, the Vickers transatlantic party consisted of ten men from the works and a specialist on Rolls Royce aero engines.

I visited my family before leaving for Newfoundland. As I was leaving for my grand adventure, Mum said, 'Please write if you can.'

'I won't write to you. I want the whole affair kept as quiet as possible. But I should like you to write to me as often as you can.'

I did write to Fred Moseley, my brother-in-law, with some instructions:

Southampton, 2 May 1919

Dear Fred,

At last, I am on way to St. Johns. The Mauretania sails tomorrow night, after being delayed so long

I want you to write to this address Mr. E. Hewer 143 Walton Rd. Woking, Surrey and ask him to send the three boxes of tools he is storing for me. I want Eddy to have them they will also be useful at home for they are both engineer's and carpenter's (have them sent passenger).

I am enclosing a few photos of my machine to be distributed amongst the family.

Well Fred I will say cheerio, will write you again.

Yours Etc.
Jack

The commander of the *Mauretania*, Captain Rostron, was a very experienced navigator, who gave Brown much good advice. The Vimy machine left later by a freight ship.

My concern, to no small degree, was that of our competitors. They all had started preparations for the Atlantic competition before we did, and they had arrived at Newfoundland before we had left England.

We were behind in the race across the Atlantic.

Chapter 15 – Newfoundland

Newfoundland, Canada, May – June 1919

WE ARRIVED AT HALIFAX on May 9th 1919. The Vimy was still at sea. After three days of ferry and train rides, we arrived at St. John's, Newfoundland on May 13th and unpacked. It had been arranged that the machine should be assembled in a cricket field at Quidi Vidi, so we took it there that same day. Screens and scaffolding were erected, and everything went well.

As Brown and I were driving back to St. John's from Ferryland, on the evening of May 18th, after an unsuccessful search for an aerodrome site, our car was held up by a motorist. 'Hawker left this afternoon,' the motorist shouted to us.

'And Raynham?' I asked.

'Machine smashed before he could get off the ground,' he answered.

We thanked him for his news and passed on to hear further details at the Cochrane Hotel in St. John's. We had rather expected the Sopwith and Martinsyde parties to make an attempt on the 18th, although the conditions were definitely unfavourable. The news of the American N.C.4's arrival at the Azores[1] had spurred them to the great adventure, despite the weather. The United States flying boats were not competing for the *Daily Mail* prize, but Hawker and Grieve wanted to gain for Great Britain the honour of being the first to cross the Atlantic by air. Harry Hawker and Kenneth Grieve made their attempt in a single engine Sopwith biplane they named the *Atlantic*.

On May 25th, Brown came to me with the news, 'I just received word about Hawker. After about 14 hours, and half-way to Ireland, the engine overheated and they were forced to change course to intercept the shipping lanes, where they were able to locate a passing freighter, the Danish *Mary*. The *Mary* did not have a functioning wireless, so it was not until the ship reached Scotland that the world knew that they were safe.' Later, Hawker and Grieve were awarded a consolation prize of £5,000 by the *Daily Mail*.

Captain Frederick Phillip Raynham and his navigator, Major Charles William Fairfax Morgan, made their attempt in a single engine Martinsyde machine they named *Raymore*. Not to lose to Hawker and Grieve, they began their flight just an hour after Hawker and Grieve had taken off. Their machine encountered a severe crosswind soon after it was airborne and slammed into the ground. Fortunately, Raynham and Morgan survived the crash.

There was a third attempt from the other side in April. Major J.C.P. Wood, the pilot, and Captain C.C. Wylie, his navigator, made an east-west attempt from England in a Short Brothers machine, the *Shamrock*, but the aircraft crashed into the Irish Sea due to a defective carburettor. They were rescued but were unable to make another attempt. Fortunately, all survived their attempts. I wondered if we would survive, too.

Our work went well until Friday, May 30th, when the weather broke. Rain poured down on us. Brown brewed tea to give us some warmth.

Bob Dicker was in charge of installing the flying and engine controls. Later that day, he reported to me, 'All the work is in the open, and the weather is making it hard going. Everyone is working up to 14 hours a day. To save time traveling in and out to St. John's, Montgomery and I are living in a couple of wing packing crates.'

'I shall make it up to you with a good breakfast each morning. You can have anything you like as long as it is bacon and eggs,' I said. Bob caught my humour, and before he could reply, I added, 'The same for your midday meal, you can have anything you like, as long as it is sausage and fried toast.'

Dicker just smiled and went on to his work.

Eventually, work on the machine had to be abandoned altogether due to the weather. Till the following Monday evening, it rained or snowed incessantly – four days of fierce gales. But even Newfoundland storms pass. The rain clouds lifted, and conditions returned to normal.

We still needed to find a place for our aerodrome. Other contestants had taken all the best sites. After much scouting of possible sites, we decided on a place called Lester's farm at Monday's Pool, where with 30 men, I blasted rocks and levelled land, walls and fences in order to get a clear run of 500 yards. I spent the first few days at the site completing the erection of the machine. There was little for Brown to do so he went to practise sending and receiving messages on his wireless. He had rigged up a receiving station on the roof of the Cochrane Hotel for that purpose.

The Rolls-Royce experts were always busy, examining and checking every little detail of their engines, so that there should be no avoidable trouble on that account. We had many watchers from St. John's, who asked many questions. They saw us filter and boil the radiator water and asked, 'What is the purpose of boiling?'

'We need to do this to remove any traces of alkali or sediment in the water which might block the engine radiators,' I answered.

We experienced rain and high wind for three days, during which the machine necessarily remained in the open, with tarpaulins covering the engines. I told one reporter, 'There is an end to most things, even to a Newfoundland storm.'

We seized our chance for a trial flight on Monday, June 9th 1919. We had meant to keep it as secret as possible, so as to avoid a crowd. It leaked out, however, and we had a large gathering in attendance. The weather was on its best behaviour, and our take-off went well. We headed over the sea for some 15 minutes. The day was clear, and the sea reflected the vivid blue of the sky. Brown spotted icebergs and whitecaps in the sea. I could see some anxiety in his eyes, as he probably realised we would be flying over a very unforgiving ocean in a few days.

Alcock (left) and Brown at Newfoundland
(From Collette Curry)

I brought the Vimy back over St. John's at an altitude of 4,000 feet. We discovered the new aerodrome, at Lester's farm, by means of a smudge-fire, lighted as a signal for us. The landing went well, however, we were heading straight for a fence on the roadside. I knew if I did not do something quickly, our adventure would be finished before we started. To save the machine from this peril, I opened up the starboard engine, which turned the machine around and away from the fence. The machine proved satisfactory, and while the wireless installation called for some attention, we felt satisfied with the result. To bolster Brown's confidence, I remarked after we landed, 'I hope to be in London before the weekend.'

Later, Brown approached me as I was looking at the controls in the Vimy. 'Do you think we have a chance at being the first across?' he asked.

'I am confident in the machine and the engines. The Rolls-Royce engines are very reliable. They are good for 100 hours; others are good for only 20. I am more concerned about the weather and our competitor in the Handley Page machine,' I answered. I was sure he wondered what we would do if we suffered the same misfortune as Hawker, and we went into the sea, so I added, 'I had one of the petrol tanks designed in the shape and function of a lifeboat, in the event we are forced down en route.'

A far more serious worry now confronted us. On examining the fuel we intended to use, a mixture of petrol and benzol, we found in it a precipitate resembling a very soft sticky resin. We could not risk such a deposit clogging our filters and perhaps stopping our fuel supply. We thought the mixture caused the precipitate, but we were not sure. Not wanting to risk an engine failure, we decided to substitute pure petrol for the mixture.

A second trial was carried out on Thursday, June 12th. This, our final test, was highly satisfactory in every way; the Rolls-Royce engines ran smoothly, and the machine behaved admirably. So pleased were we with the result that we decided to set off for Ireland as quickly as possible. On Friday, the tanks were filled with petrol, lubricating oil, and water. A provision for an unexpected landing was a store cupboard fitted in the tail, where the emergency rations were placed. These would be clear of the waves when the machine floated.

On the evening of Friday the 13th, Brown informed me, 'The last coat of dope has dried and nothing has been overlooked.' The dope applied to the cotton fabric on the wings and fuselage provided tautening of the material.

'All looks well; let us plan on leaving early tomorrow morning,' I said, somewhat disappointed we would not leave on the 13th, my lucky number.

Before we retired, a young reporter approached me and asked, 'How will you find your way across the ocean?'

I motioned for him to come to me, and I whispered with a straight face, 'I've got six carrier pigeons from England hidden away. After we've flown 300 miles, I plan to release one pigeon, which of course would immediately head for its home in England. Naturally, I'll follow the bird in the plane. I'll release one bird every 300 miles, with the last bird taking us over Ireland.'

I later learnt that my little joke was published as an exclusive.

We went to bed about 7:00 p.m., while the mechanics remained all night with the machine. At 3:30 a.m., one of the mechanics called us, and we joined them and others at the aerodrome, where the machine had been moved into position for take-off.

'Strong westerly wind. Conditions otherwise favourable,' reported the meteorological officer at 4:00 a.m.

The strong crosswind remained until later in the day. Finally, there was some slackening of the wind, so Brown and I decided to 'chance our arms.'

'We shall land at the Galway coast,' Brown stated, very confident of his navigation skills.

I laughed and said, 'Yes, we shall hang our hats on the aerials of the Clifden wireless station as we go by!'

Everything was ready for take-off. I was concerned, as the machine had never been flown fully loaded. I beckoned Bob Dicker and said, 'Come on Bob, get those engines started. It's time to go.'

My request was promptly carried out and I heard the sweet sound of those engines. I ran the engines up and throttled back to warm them up. When everything was satisfactory, I disconnected the starting magnetos and engine switches, to avoid stoppage due to possible short circuits, and signalled for the wheel chocks to be pulled away. With throttles open, and the engines all out, the Vimy lumbered into the westerly wind. It was at 4:13 p.m. (Greenwich time) that we began the long-hoped-for take-off.

We gained speed, but the Vimy was still firmly on the ground at about 100 yards from the end of the strip, heading straight for a low thick stone wall. Suddenly, the Vimy rose about five feet, dropped, and then bumped over the wall with about two feet to spare. Once over the wall, the ground immediately fell away. We were on our way.

Before we left Newfoundland, however, we were informed that Admiral Mark Kerr's four-engine V/1500 Handley Page had finished the trial flights at the same time we had and was being prepared for flight. Thus, it became a competition with the Handley Page machine and with the elements above the north Atlantic.

Chapter 16 – Conquering the Atlantic

Newfoundland, Canada and Clifden, Ireland, June 1919

OUR TAKE-OFF was up a slight gradient, and there was sufficient room to start heading into the wind. We managed well, however, but a high range of hills on either side made climbing difficult and occasional bumps kept us down. Several times I held my breath, from fear that our undercarriage would hit a roof or a treetop. I could feel the perspiration streaming down my face from the anxiety.

Emerging from the valley, however, we got head to wind, and having climbed quickly to 1,000 feet, we set out on the eventful journey. Rising steadily, we passed over St. John's and Cabot Hill, and at 4:28 p.m. (Greenwich time) we crossed the shore bound S.E. 124° (magnetic); destination Ireland.

For a time, Brown and I attempted short conversations through the telephone. Its earpieces were under our fur caps, and round our necks were sensitive receivers for transmitting the throat vibrations that accompany speech. At about 4:45 p.m., I removed the earpieces because they were too painful, and for the rest of the flight we communicated by signs and notes.

Brown informed me of a course change by passing over a scrap of paper from his notebook, which read, 'Keep her nearer 120 than 140.' A few minutes later he wrote, 'Wireless generator smashed. The propeller has gone.[1] '

Visibility was good at the outset, but I saw ahead the dreaded fog of Newfoundland, and within a short space we were racing between it and the clouds. At 5:45 p.m., Brown scribbled, 'I can't get an obs. in this fog. Will estimate that same wind holds and work by dead reckoning.'

The fog increased until presently we found ourselves buried in heavy fog. I held onto the control wheel and kept my eyes on the compass and invisible horizon.[2] The early evening was not dull. Just after 6:00 p.m., the starboard engine startled us with loud noises like that of a machine gun. I thought of the engine trouble Hawker and Grieve had that caused them to descend into the mid-Atlantic. We both looked at the engine and saw that a piece of exhaust pipe had split away and was shaking before the rush of air. Eventually, it became white hot, broke off, and was blown away. The result was three cylinders were exhausting straight into the air and causing a loud noise.

At 7:40 p.m., Brown handed me another note, 'If you get above clouds, we will get a good fix tonight and hope for clear weather tomorrow. Not at any risky expense to engine though. We have four hours yet to climb.'

At 9:15 p.m. we caught sight of the sea for a few minutes. Brown wrote another note to me, 'Through a rather bad patch, I have just made our ground speed 140 knots, and from the sun's altitude, we must be much farther east and south than I calculated.'

At 9:30 p.m., when the light was fading, he scribbled an enquiry to me, 'Can you get above these clouds at say, 60 deg. We must get stars as soon as poss.'

I nodded and proceeded to climb as steeply as I dared.

The conditions were bad, but I held on to the course set by Brown till suddenly, at 12:25 a.m., we shot into a clear patch in the clouds. I signalled to Brown to take a reading.

Quickly, Brown fixed our position from the Pole Star, Vega and the moon, and he made me aware of our position, 'We have flown 850 nautical miles at an average speed of 106 knots. We are slightly south of the correct course.' Being very meticulous, Brown made note of this and everything else in his navigation log.

Outside the cockpit was bitter cold, but inside it was a sheltered warmth, due to the protective windscreen, the heat from the radiator, and our thick flying suits. Our only discomfort was the impossibility of making any but cramped movements.

For about half an hour, we had a clear sky overhead. Then, however, thick fog enveloped us again, and a second mishap nearly brought our venture to an untimely end.

The air-speed indicator failed to register, and bad bumps prevented me from holding to our course. From side to side rocked the machine, and it was hard to know in what position we really were. A spin was the inevitable result. From an altitude of 4,000 feet, we twirled rapidly downward.

It was a tense moment for us, and when at last we emerged from the fog we were close down over the water at an extremely dangerous angle. The whitecaps were rolling along too close to be comfortable, but a quick glimpse of the horizon enabled me to regain control of the machine at 50 feet above the waves.

Getting back on course, I climbed to 7,000 feet above sea level. Occasionally, we saw the moon. I indicated to Brown to fix our position.

'I cannot,' Brown indicated.

I hoped to get above the clouds, but it seemed impossible. Climb as we might, the clouds seemed to climb still faster. For five or six hours, we continued the attempt, but in vain, with hail and sleet lashing across us.

At 8,800 feet, we discovered that the glass face of the petrol overflow gauge, which showed whether or not the supply of fuel for the engines was correct, had become obscured by snow. To guard against carburettor trouble, it was necessary that I should be able to read the gauge.

Brown saw the problem, too and signalled to me what he was about to attempt. He stood up in the open cockpit on my left, leaned on top of the fuselage, while holding a strut for maintenance of balance. The gauge, fixed on one of the centre section struts, was within his reach.

I was worried that the violent rush of air would push him back and off the machine, or I would be unable to hold the machine steady to keep him aboard. During this time, once again, I felt the perspiration on my face from the anxiety I experienced. Brown, however, was able to successfully clear the snow from the gauge. Unfortunately for him, this procedure needed to be repeated several more times in our flight.[3]

I kept climbing to get above the sleet and snow. The top sides of the machine were covered completely by a crust of frozen sleet. The sleet imbedded itself in the hinges of the ailerons and jammed them, so we had scarcely any lateral stability. Fortunately, the Vimy possesses plenty of inherent lateral stability, and as the rudder controls were never clogged by the sleet, I was able to hold to the right direction.

At 5:00 a.m., we reached 11,000 feet and were met with some reward, as the sun at last appeared trying to force its way through the clouds. Brown was definitely worried about our situation. He wrote another note and handed it to me, 'We had better go lower, where the air is warmer, and where we might pick up a steamer.'

We started our descent, but immediately the starboard engine began to pop ominously, as if it were backfiring through one of its carburettors. I throttled back, while keeping the machine in a slow glide. The pops ceased.

By 8:00 a.m. we were at 1,000 feet, and close over the sea. The atmosphere was much warmer, and the ailerons were operating again. We found a strong sou'wester blowing, but we were still in a cloud.

I continued to descend at a very shallow angle, not knowing if the cloud extended to the ocean, nor at what moment the undercarriage might catch the waves. I loosened my safety belt and was ready to abandon ship if we hit the water. I was worried that the change in barometric pressure from the descent could cause the aneroid[4] to show a false altitude reading.

But once again we were lucky. At 500 feet altitude, the Vimy emerged from the cloud, and we saw the ocean. I opened up the engines and they performed flawlessly. Brown started some navigational calculations and handed me a note, 'Don't be afraid of going S. We have too much N. already.' The note had a sketch showing his recommendation of 170°.

We decided to stay at the low altitude and skimmed swiftly over the surface of the water for some 40 minutes. Suddenly, and without a moment's warning, so fast were we traveling, the sea disappeared from beneath, and the shimmering green of land appeared and gladdened our hearts.

But it was only for a moment. Again, we were over the western ocean, and we learnt later that we had crossed the outlying islands of Eashal and Turbot. In the rain, we could see no mainland, but we knew it must be near at hand, and later we were not surprised but delighted to find ourselves over it.

Pleased with our position, we flew on for several minutes till suddenly the tall masts of a wireless station stood before us. (It was Clifden we learnt afterwards.) Firing Verey signals, we circled round the station and over Clifden town, but without response. By the wireless station, I had noticed what appeared to be a suitable field, so we decided to land there.

What I had thought was a field turned out to be a bog. Still, we received no hurt, although the machine was somewhat shaken by the sudden stoppage. The front of it was bent in some places and smashed in others. The four propeller blades were buried in the ground, although none was broken. The petrol connections had snapped, and petrol began to fill the rear cockpit. Fortunately, I had switched off the current to the magnetos before landing, and the petrol did not ignite.

We scrambled out as best we could and salved the mailbag and our instruments. The petrol rising rapidly had damaged Brown's chart and his Baker navigating machine, before we could withdraw them.

Brown fired two Verey flares, as a signal for help. Almost immediately a small party of officers and men belonging to the military detachment at Clifden approached from the wireless station.

'Anybody hurt?' they asked when they were within shouting distance.

'No,' both Brown and I answered.

'Where are you from?' they asked.

'America,' I yelled back. 'Yesterday we were in America. We're Alcock and Brown.' I showed them the white sealed airmail bag from St. John's.

Landing in a peat bog at Clifden, Ireland
(From Collette Curry)

They had no idea who we were, and on hearing that we had crossed the Atlantic their astonishment was exceeded only by their enthusiasm. We arrived at Clifden at 8:25 a.m., so the time occupied in the journey was 15 hours 57 minutes.[5] As the distance covered was 1,960 miles, the average speed was over two miles a minute, which must be admitted, was very good going in the circumstances. We were too tired to think much of the feat at the time, and our subsequent experiences were of an even more strenuous but gratifying nature.

With our wireless utterly useless during the flight, we still wondered what had happened to the Handley Page. Did Admiral Kerr cross before us?

Chapter 17 – The Reception

Ireland and England, June 1919

THE WALK ACROSS DERRYGIMLA BOG was a dragging discomfort, as we were burdened with our flying suits and heavy boots. Along the way, Brown turned to me and asked, 'What do you think of that for fancy navigating?'

'Very good,' I replied and shook hands with Brown.

At the wireless station, we received every kindness. When we recovered our hearing, we asked, 'What have you heard about the Handley Page?'

'We have heard nothing about either its departure or arrival,' several answered.

'Maybe they have flown direct to England?' Brown reasoned.

We were not sure about that, as no one at the wireless station had heard anything. Later in the day, we heard that the Handley Page had radiator problems, which grounded it. We had won the *Daily Mail* prise!

Our first act was to send telegrams to the *Daily Mail*, Vickers, and the Royal Aero Club. Remembering our crew waiting anxiously in St. John's, we dispatched the first message, 'Your hard work and splendid efforts have been amply rewarded. We did not let you down. Alcock and Brown.'

The news of our arrival was soon going out in every direction. During this exchange, we quickly devoured some bacon and eggs and washed down generous mugs of tea.

At 12:55 p.m., Brown sent a cable to Kathleen Kennedy: 'Landed Clifden, Ireland safely this morning. Will be with you soon. Love Teddy.'

We cabled the Rolls-Royce managing director, Claude Johnson: 'Congratulations on performance of the two Eagle Rolls-Royce engines which propelled the Vickers Vimy safely across the Atlantic.'

J.L. Hodsen of the *Daily Mail* arrived with a photographer. He was the first of the newspapermen to interview us. Hodsen asked how we were feeling.

'We didn't do so badly, did we?' Brown responded.

'I am not at all tired, but I would like to stand up for the rest of my life,' I answered.

'I am a bit fagged out,' Brown added.

The newspapermen continued to ask about our flight.

'I got more fun out of my first solo,' answered Brown.

'This was a job of work. It was a difficult task, but we are too near it to realise what we have done. We have had a terrible journey. The wonder is we are here at all. We scarcely saw the sun or the moon or the stars. For hours we saw none of them. The fog was very dense, and at times we had to descend to within 300 feet of the sea,' I answered and then gave them more details of the flight.

'How did you feel about flying over the Atlantic during the night?' another asked.

'I had done a considerable amount of night flying in the past, and the sense of loneliness that might be supposed to accompany it had long since worn off. Indeed, I do not think that either of us had any thought of what we were flying over, but merely intent on getting the machine across. There was no sense of remoteness; we were just too keen on our work,' I replied.

After a while, one newsman asked me about the future of transatlantic flight.

'The flight has shown that the Atlantic is practicable, but I think it should be done not with an aeroplane or a seaplane, but with a flying boat,' I noted.

'What do you think of your aeroplane?' another correspondent asked.

'The engines are Rolls-Royces, and they ran perfectly all through. We did not lose a spoonful of water on the journey and no petrol. When I landed, I had only two-thirds of the supply exhausted,' I noted, then took my leisure with a cigarette, which I had not enjoyed for some time for obvious reasons.

We received more congratulations, and I said to one of the correspondents, 'Yes, I'm glad we did it.'

Shortly after 4:00 p.m., we left the Marconi station for Galway, which was about 50 miles away. For most of the journey, Brown and I remained silent, as we were recovering in our minds from all the excitement of the past two days. Our arrival had spread all over the districts between Clifden and Galway, and curious crowds gathered along the way. Near Galway, our motorcar was stopped by another motorcar with Major Mays of the Royal Aero Club. He was on his way to Clifden to inspect the seals on the Vimy. Mays was to make sure that we had not landed in Ireland in a machine different than the one in which we had left Newfoundland.

At Galway, our hosts, realising how tired we must be, made it a short and informal reception. Afterwards, we slept for the first time in more than 40 hours.

Our physical systems had become adjusted to the clocks of Newfoundland. We experienced difficulty rising at 7:00 a.m., for our bodies suggested that it was 3:30 a.m. This lasted several days. Probably, all passengers traveling the same way in the future will experience this feeling.

'NEWFOUNDLAND TO IRELAND UNDER 16 HOURS. ALL BRITISH TRIUMPH,' read a headline that greeted us on Monday, June 16th 1919.

The *Times* carried a full report on the flight with its congratulations.

In Paris, *Matin*, provided further acclaim, 'The achievement is more than remarkable. It is sensational.' *Le Journal* made similar remarks.

In the U.S.A., *The World* and *The New York Times* covered our flight.

One newspaper reporter had quoted Brown as saying, 'Right from the beginning, when we first met, we worked together wonderfully. To my mind, Alcock has the ideal temperament for flying. Nothing seems to upset him, and he does not appear to have any nerves. He is always bubbling over with good humour. The way he nursed the engines on the flight was great.'

The same reporter quoted my comment on Brown, 'I consider that the navigation on the trip under the trying circumstances and bad visibility was really wonderful.'

U.S. President Wilson sent his congratulations. Glenn Curtiss, a prominent flyer in America, did the same.

Our Prime Minister, Lloyd George, wrote to me: 'Heartiest congratulations to you and Lt. Brown on your audacious and splendid flight. It is a splendid achievement. I am especially delighted the two British officers, who fought in the war, should have been the first to Europe from America in a non-stop flight.'

At the hotel in Galway at 1:00 p.m., town officials assembled. Thomas Macdonagh delivered the first address. He opened with, 'On behalf of the citizens of Galway, which has the honour of being the landing place of the first non-stop transatlantic flight, the Galway County Council heartily welcomes the intrepid airmen who successfully carried out this magnificent feat.'

After Capt. Redengton read the urban council's contribution, I was asked to make a speech. After I arose for my first public speech, I said, 'I hope you will excuse me. I am not used to this sort of thing. I wish to express to you our feelings of deep gratitude, and if you will kindly excuse me.' I quickly sat down.

We were then presented with the famous Claddagh rings, which date back several centuries, and were also given to Queen Victoria and King Edward VII, when they visited Galway in 1903.

As we tried on the rings, there was a call, 'Three cheers,' and all the officials joined in the singing of 'For They Are Jolly Good Fellows.'

'I shall always keep my ring and wear it as a lucky charm,' I told the well-wishers.

After more celebration, we were taken outside, where we were met by a large crowd. Lord Killanin and Major Mayo escorted us to the 2:25 p.m. Dublin train. On the journey to Dublin, crowds gathered along the way, even at stations where the train did not stop. When we did stop, people constantly asked for our autographs. We signed our names hundreds of times.

At one stop, a small lad presented me with a model aeroplane he had made.

I was touched by the boy and said to him, 'Thanks very much indeed old chap, a jolly good bit of work. Very much like a B.E. tractor biplane, isn't it?'

He seemed pleased with my comment, and then I was ushered onto the train once again.

We arrived at Dublin at 6:30 p.m., and students from Trinity College promptly kidnapped me. The provost finally rescued me, and I joined Brown, who said he was still tired.

'If I am due for a reaction from the strain and excitement of the flight, it is a long time coming,' I said.

Under special guard, we were taken to the lodge of the Chief Secretary, Mr. Macpherson. There, we spent our last night in Ireland.

The morning of Tuesday June 17th, we embarked on our return home in the SS *Ulster* for Kingstown.[1] Some of the passengers recognised us, and one asked me, 'What do you think of Ireland?'

Royal Aero Club Award
(From Tony Alcock)

'I want to thank the Irish people for the royal reception they have given to Lieutenant Brown and myself on our way from Galway to Dublin. It was really wonderful. This reception is really harder than the actual flight and entails a bigger strain. I was nearly torn to pieces yesterday!' I answered.

When we disembarked at Holyhead, we were met by Capt. Vickers, designer Rex Pierson, and Lord Northcliff's friend, Claude Johnson, who was general manager of Rolls-Royce. Once again, we were met by crowds at nearly every railway station along the way to London. Kathleen Kennedy met Brown at Rugby. At almost every stop, we were forced to make brief speeches.

On our way to London, we met up with Harry Hawker, who joined our procession. Thousands of people lined the streets along our route in London as we made our way to the Royal Aero Club. Vice-chairman General Holden welcomed us, and I handed him his white bag of mail from Newfoundland. I jokingly said, 'I am the first transatlantic postman.' Then he presented me with a trophy to commemorate our flight.

On Friday June 20th, there was a luncheon at the Savoy, where Winston Churchill, the Secretary of State for War, presented the *Daily Mail's* £10,000 check to us. Mr. Churchill informed us that King George V had awarded both of us Knight Commandership of the Order of the British Empire. He told us to go to Windsor the next day to receive the accolade from the king.

On Saturday afternoon, we went there by train, and were met by the Royal Carriage, which was drawn by a pair of greys. There was great ovation as we went along. At the Royal Castle, His Majesty congratulated us for our flight, and we had an interesting chat with him and other members of the royal family.

It was only after these events, when in some possession of some leisure, that any retrospect was possible. First, one must give due praise to the machine, which reflects great credit on the makers, Messrs. Vickers, and their designer, R.K. Pierson. It behaved splendidly throughout, and neither figuratively nor literally let us down once. And the Roll-Royce engine, a wonderful piece of work, ran without a hitch throughout. In the varying climatic conditions of the voyage, the Claudel-Hobson carburettor deserves praise also, as does the All-British Warford Magneto, which worked admirably in the damp atmosphere encountered. There remains the fuel, the Shell spirit, and of it I cannot speak too highly. Like all else, it went to make the happy combination, which ensured success.[2]

Vickers offered both Brown and me jobs for life, and I was looking forward to continuing my association with Vickers.

Chapter 18 – New Ventures

England, July – December 1919

IN THE FOLLOWING MONTHS, we endured various sittings for official portraits. We were pressured to write accounts of our flight, and I gave a brief summary, which was written up and published by *Badminton Magazine*. Brown worked with journalist Alan Bott on a more elaborate version of our flight.

In July, I was presented a certificate from the City of Manchester for our transatlantic flight.

On July 29th 1919, Brown married Kathleen Kennedy at the Chapel Royal, Savoy, off London's Strand. They went to America on their honeymoon aboard the *Mauretania*, the same ship we had taken to Newfoundland three months earlier. I had hoped to arrive at their wedding unnoticed, but the cheering spectators embarrassed me.

I was happy to be back at Brooklands, testing machines for Vickers. Five weeks after the flight, I met Bob Dicker at Brooklands alongside the track.

'Now that you are home, how do you feel about the flight?' he asked me.

'It was a difficult task. It is a wonder that we made it with the weather as it was. Brown certainly performed well in many respects,' I answered.

'Do you have any regrets?' Dicker probed.

'My biggest disappointment was our landing in Clifden. It was a shame the amount of damage to the Vimy.'

I saw Bob Dicker several times after that and he agreed to help me start a motor agency. I had several good makers to back me, and I knew that Bob would be very helpful. He was now selling motorcycles and was experienced in starting a business.

In August, I accompanied Capt. Cockerell, who flew a Vimy Commercial from England to Eerste Luchtverkeer Tentoonstelling Amsterdam, E.L.T.A. (First Aviation Exhibition Amsterdam) for an aviation exhibition. We had a large party of passengers with their luggage on board. One passenger, Mr. Mullens, filmed the journey and took an excellent pictures of the start, the journey itself, 'life on board' and the arrival at E.L.T.A. This film, later shown at the Cinema de Munt where it was greatly appreciated.

Vickers Vimy Commercial, 1919
(From the Public Domain)

A most unusual event occurred recently, which related to my last day at Mudros during the Great War. After my battle with the German machines, the pilot and some parts from one of the machines was picked up by the Royal Navy. I just received a letter from the Bulgarian lady relating to this incident:

Dear Sir,

I apologise for having taken the liberty to write you without being known by you---the liberty to write you and to have your attention just for a moment follow these lines.

Therefore, I must introduce myself to you. I am the young lady to whom the German pilot Muller, that you brought down two years ago, had written his last greetings on the propeller of the wrecked seaplane. Some time ago I received the number of the Daily Mirror which contains a picture of the propeller and your photo. I was very curious to know something about you, and Captain Wildish and Commander Greig, who know you very well, told me what a good flyer you are, and that just recently you gained the prize in a concourse for a flight over the ocean. I congratulate you for that.

My friend Muller is a very good pilot, he was one of the best, and I see that he has had a worthy ending, more than that, you must have excelled him to have beaten him down.

I myself have never been in an aeroplane, and don't know what the sensations may be when one is so high up, but there must be something that makes you risk your life so readily, you really must feel great having conquered the space.

I began by introducing myself, and I must say something more to let you have a complete picture. I am a Bulgarian and am a graduate of an American college; I lived several years among English and American people and learned to love your language and literature. Last winter we had great many English people here, and I had the chance to practice the language, and enjoyed it greatly.

I shall be immensely pleased if you send me a good photograph of yourself, and if you add some snapshots of your seaplane too, I shall enjoy them none the less, and with the picture in the Daily Mirror I shall highly treasure. I shall not say how I would wish to possess the famous propeller, but rightly it belongs to you; it is your trophy, for it is you have risked your life.

Believing you will answer my small request,

I am with very best wishes,

Yours sincerely,

(sd.) STRAOKA ILIEVNA.

Varna, Bulgaria,
21 Vornishka.
1 August 1919

The lady mistakenly thought the message was on a propeller. Actually, it was written on part of a seaplane float, which the Royal Navy picked up along with the pilot. The float was given to me after the war, as a souvenir. I was not disposed to establishing a relationship of this sort, so I postponed a response to the lady's request until after I had some time to consider a fitting answer. At present, I am totally immersed in a new venture with Messrs. Vickers.

I celebrated my twenty-seventh birthday on November 6th 1919. A week later, the Smith brothers, Ross and Keith, left Hounslow[1] on the first-ever flight to Australia. Four Australian fliers were lost in the race, which had 18 entrants.

On December 15th, Vickers held a public celebration for the Smith brothers, who had completed the first England-Australia flight three days earlier. Douglas Vickers, Moore-Brabazon and Claude Johnson of Rolls-Royce attended. I was there for the presentation of our Vimy to the London Science Museum.

'The curio-hunter was there and made bounty of the machine[2] – so that this is not the only place where relics of the Atlantic machine would be shown,' Vickers warned.

'I can vouch for that!' I added, midst the laughter.

Returning to Brooklands, I planned on flying the new single-engine Vickers Viking Mark I amphibian biplane to the aeronautical exhibition in Paris. I invited Bob Dicker to accompany me, but he was busy with his motorcycle business and declined my offer.

Epilogue

SIR JOHN ALCOCK, on December 18th 1919, arrived early at Brooklands from his home at Mayfield Road, Weybridge. The weather was miserable. There were low clouds and wind-driven rain. Several of his colleagues suggested that he postpone the trip, particularly as he was flying without a navigator. But the test pilot for the Vickers Viking felt he had already waited too long, as the Paris exhibition was to open the following day.

He encountered blustery conditions over the Channel but reached the Normandy coast safely. He most likely was following the Seine when a heavy fog reduced his visibility to 100 yards. When he descended to get his direction, his aircraft struck a tree at Cote d'Evrard, which was about 25 miles north of Rouen. He sustained a skull fracture and was taken to a hospital at Rouen, but he never regained consciousness. He died that same day,

John Alcock in Uniform
(From Tony Alcock)

December 18th, at 4:00 p.m. He was 27 years of age.

Sir John William Alcock was buried in Southern Cemetery, Manchester, England. One inscription on his tombstone read, 'In memory of our dear brother whose soul made the great flight 18th December 1919.'

<>\<>\<>

ARTHUR BROWN was in San Francisco when Alcock had his accident and was shocked when he heard the news. Reporters heard him say, 'A true sacrifice for humanity.' He later went to work for Metropolitan-Vickers and was involved with the sale of Vimys to China.

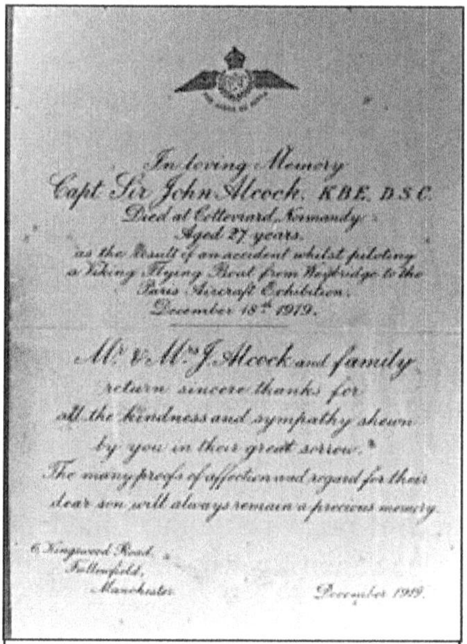

In loving Memory
Capt Sir John Alcock. KBE. DSC.
Died at Cottevrard, Normandy
Aged 27 years.
as the Result of an accident whilst piloting
a Viking Flying Boat from Weybridge to the
Paris Aircraft Exhibition.
December 18th 1919.

Mr & Mrs J. Alcock and family
return sincere thanks for
all the kindness and sympathy shewn
by you in their great sorrow.
The many proofs of affection and regard for their
dear son will always remain a precious memory.

6 Kingswood Road,
Fallowfield,
Manchester.

December 1919.

Thank You Note
(From MOSI, Ref. 1997.36.3)

During the Second World War, he re-joined the Royal Air Force to train pilots in navigation and engineering. His only son was killed in the war. On October 24th 1948, he died at his home, Belgrave Court, in the Uplands district of Swansea, of an accidental overdose of Veronal.[1] He was 62 years of age.

<>\<>\<>

MAJOR GEORGE HERBERT SCOTT, CBE, RAF, with his crew and passengers on July 2nd 1919, flew the airship (dirigible) *R34* from East Fortune, Scotland to Newfoundland, Canada to Mineola, Long Island, New York, then made the return trip to Pulham, England. It flew about 7,000 miles in 183 hours 15 minutes, thus completing the first-round trip across the Atlantic. Airships were considered the future of transatlantic travel in 1919 until the loss of the *R101* in 1930 and the loss of the *Hindenburg* in 1936.

<><><>

THE VICKERS VIMY reached full-service status in July 1919, when it entered service with 58 Squadron in Egypt. The aircraft formed the main heavy bomber force of the R.A.F. for much of the 1920s, where it served as a front-line bomber in the Middle East and in the United Kingdom. It was replaced by the Vickers Virginia but continued to equip Special Reserve 502 Squadron in Northern Ireland until 1929.

The Vimy continued in use as a training aircraft, many being re-engined with Bristol Jupiter or Armstrong Siddley Jaguar radial engines. The *Atlantic Vimy*, flown by Alcock and Brown, is preserved at the London Science Museum.

<><><>

CHARLES LINDBERG made the first non-stop solo flight across the Atlantic in the *Spirit of St. Louis* on May 20th and 21st 1927, eight years after Alcock and Brown conquered the Atlantic. He left Roosevelt Field, New York, to stay aloft for 33 hours 39 minutes and travelled 3,600 miles to Le Bourget Field. Upon landing in Paris, after his own record-breaking flight, Lindbergh told the crowd welcoming him, 'Alcock and Brown showed me the way!'

<><><>

LOUIS BLÉRIOT formed his own company after the war, Bleriot Aeronautique, for the development of commercial aircraft, which was more successful than the company founded by the Wright Brothers.

In 1927, Blériot was present to welcome Charles Lindbergh when he landed at Le Bourget Field, outside Paris, after his transatlantic flight. The two men, separated in age by 30 years, each had made history by crossing famous bodies of water. Together they participated in a famous photo opportunity in Paris. Bleriot died of a heart attack on August 2nd 1936. He was 64 years of age.

<><><>

THOMAS SOPWITH, after the transatlantic attempt and voluntary liquidation of his firm, started a new company named after his chief engineer and test pilot, Harry Hawker. Hawker was chairman of Hawker Aircraft, though Sopwith ran the company.

Hawker Aircraft later acquired the companies of J.D. Siddeley and became Hawker Siddeley, which produced the famous Hurricane fighter aircraft of World War II. Avro, a part of the Hawker Siddeley group, produced the famous Lancaster bomber of that same war.

Sopwith continued to work as a consultant until 1980. On his 100th birthday, military aircraft from many different eras performed a flyover above his home. He died on January 27th 1989, aged 101.

<><><>

HARRY HAWKER was flying a Nieuport Goshawk at Hendon on July 12th 1921, when he lost control of the plane from paralysis caused by a sudden haemorrhage of an abscess on the spine. The machine caught fire on impact and Hawker died a few minutes later. He was 32 years of age.

<><><>

FREDDIE RAYNHAM later formed the Aircraft Survey Co. and the Indian Air Survey and Transport Co. with Ronald Kemp. He was managing director of the Indian concern and flew extensively in India and the Far East.

During the Second World War he was a member of the Air Ministry's Air Accident Board. He died in 1954, aged 60, while on a six-year tour of the United States with his wife. He was buried in The Evergreen Cemetery in Colorado Springs, USA.

<><><>

AMELIA EARHART, in May 1932, became the first woman to fly solo across the Atlantic. She flew a Pratt & Whitney Wasp-powered Lockheed Vega. Her flight was from Harbour Grace, Newfoundland, to Ireland in approximately 15 hours.

<><><>

EDWARD SAMSON JOHN ALCOCK, known as Captain John Alcock, was a distinguished flyer like his elder brother, Jack. In 1955, he held the world record for flying hours as a pilot culminating some 24,500 hours, representing a distance of some 5,500,000 miles in 40 years of service. Initially serving with the RAF, he flew fighters and bombers and later became a flying instructor.

In 1929, he was ordered by the Air Ministry together with eleven other RAF pilots, to join Imperial Airways (later British Overseas Airways Corporation, BOAC), as the government wanted to expand civil aviation. In so doing, he became involved in establishing passenger air routes to Egypt, India and South Africa and mail routes to Australia.

In 1936, he converted onto Flying Boats and established flying boat routes to the Middle and Far East. During WW II, he flew unarmed flying boats in support of operations and, in 1943, was awarded the Kings Commendation for services in the air for the evacuation of British personnel from Crete.

After the war, he was sent to run the RAF Heavy Transport Conversion Unit, No. 1384, at Ossington, Nottingham. In 1946, he went back to BOAC to be the Chief Flying Instructor at the BOAC Airways Training School at Aldermaston. He retired from BOAC in 1950 but continued to fly or be involved in aviation until 1955.

<><><>

A J H "TONY" ALCOCK, Edward's son, also pursued a career in aviation and joined the RAF and flew primarily Lightning and Phantom fighters but also flew other fighters and helicopters. Tony retired as a Group Captain after 35 years in the RAF: notable duties included Commander of British Forces operating in Turkey in support of air policing over Iraq, Commander of RAF forces and Deputy Commander of British Forces in the Falklands, Head of the NATO Tactical Evaluation in Europe, pilot exchange officer duties with the USAF serving at Elmendorf AB, Alaska, flying the Phantom F4E and Commanding Officer of RAF Wattisham, RAF Mount Pleasant and 29 (F) Squadron.

He was awarded an MBE[2] for his work with the Central Tactics and Trials Organisation where he wrote the Phantom Tactics Manual when the Phantom became the RAF's primary Air Defence aircraft, conducted a number of flying trials and produced an Electronics Counter Measure training film for aircrews.

<><><>

NEVILLE ALCOCK has lived in Manchester all his life and following the death of his father, Albert, a younger brother of Sir John Alcock, Neville has been the family's point of contact for Alcock and Brown. Over the years, he has acted as the de facto guardian of the Alcock memorabilia and has ensured that the items held at the Manchester Town Hall, the Museum of Science and Industry (MOSI) and the commemorative statues and Blue Plaques of historical interest are properly maintained and appropriately displayed. In addition, Neville coordinates the publication of important anniversary dates in the local press.

<><><>

A STATUE OF ALCOCK AND BROWN was erected at London's Heathrow Airport In 1954.

Statue of Alcock and Brown
Heathrow Airport
(From the Public Domain)

Bibliography

Articles:

Collette Curry, "A Manchester Lad," *FlyPast*, November and December 1992.

"Blackpool Flying Meeting 18th/23rd October 1909," (notes from the programme for the meeting in 1909, bought by Mr. Reed.)

"British Notes this Week," *Flight International*, 1910-1914.

"Brooklands," *Flight International*, November 29, 1913.

"Crossing the Big Puddle," *Boys' Life*, June 1993.

"Liquidation of Empress Motor Car and Aviation Company," *The London Gazette*, September 7, 1915.

"F.P. Raynham," *Flight International*, May 21, 1954.

"From the British Flying Grounds," *Flight International*, 1910-1914.

Maurice Ducrocq, "How I Learnt to Fly," *Flight International*, November 5, 1910.

"A.D George, "John Alcock – The Early Years," *Manchester Memoirs*, Vol. 124, (Manchester, England: Manchester Literary & Philosophical Publications Ltd, 1986) pg. 131.

"John Alcock's Service Record," (notes from Fleet Air Arm Museum).

"London-Manchester-London Air Race," *Flight International,* June 26, 1914.

"Louis Bleriot's Record-setting Flight Across the English Channel," *U.S. Centennial of Flight Commission,* U.S. Government.

"Parsons' Farman Biplane with Stability Control," *Flight International*, May 17, June 21, August 2 and 30, September 20, 1913.

"Parsons' Sommer Biplane," *Flight International*, December 14 and 21, 1912.

"Robert E. Dicker," Interview, 17 March 1977.

"Sydney Camm's Recollections – Raynham the Unlucky," *Flight International,* June 18, 1954.

"The New Daily Mail Prizes," *Flight International*, April 5, 1913.

Henry Woodhouse, "The Trans-Atlantic Flights," *Flight International*, June 1919.

"Wood, Ocean Flier, Drops into the Irish Sea," *New York Times*, April 20, 1919.

Books:

Sir John Alcock and Sir Arthur Whitten Brown, *Our Transatlantic Flight* (London: William Kimber, 1969).

Chris Aspin, *Dizzy Heights: The Story of Lancashire's First Flying Men* (Helmshore Local History Society, 1988).

J. M. Bruce, *The Aeroplanes of the Royal Flying Corps (Military Wing)* (London: Putnam, 1982)

David Culshaw and Peter Horrobin, *The Complete Catalogue of British Cars 1895-1975.* (Dorchester: Veloce Publishing plc. 1999).

Michael H Goodall and Albert E. Tagg, *British Aircraft before the Great War* (Atglen, PA: Schiffer Publishing Co., 2001).

Ann Hagedorn, *Savage Peace: Hope and Fear in America, 1919* (New York: Simon & Shuster, 2007) 235.

A.F.C. Hillstead, *Fifty Years with Motor Cars* (London: Faber and Faber, 1960)

A. J. Jackson, *Avro Aircraft Since 1908* (London: Putnam, 1990)

Peter Lewis, *British Aircraft 1809 (sic)-1914* (London: Putnam, 1962)

Brendan Lynch, *Yesterday We Were in America* (Newbury Park, California: Haynes North America Inc.).

Brian R. Robinson, *Aviation in Manchester, A Short History* (Leicester, Great Britain: A.B. Printers for the Manchester Branch of the Royal Aeronautical Society, 1977)

Ross Smyth, *The Lindbergh of Canada: The Erroll Boyd Story* (Burnstown, Ontario: General Store Publishing House, 1997)

Henry Serrano Villard, *Contact! The Story of the Early Aviators* (Mineola, NY: Dover Publications, 2002)

Graham Wallace, *The Flight of Alcock and Brown* (London: Putnam, 1955)

Sir Leonard Woolley, *From Kastamuni to Kedos: Being a Record of Experiences of Prisoners of War* (Oxford: Basil Blackwell, 1921).

Websites:

"Captain Jack Alcock," Collections Department, Museum of Science and Industry accessed 2012, www.mosi.org.uk/media/33871083 /captainjackalcock.pdf.

"Charles Rolls, 1877-1910," accessed 2012, www.earlyaviators.com /erolls.htm.

"ELTA – The First Aviation Exhibition 1919," Rob J.M. Mulder, accessed 2012, www.CrossandCockade,com

"Harry Hawker: The Boy who wanted to Fly," Kingston Historical Website, accessed 2012, http://localhistory.kingston.vic.gov.au/htm /article/188.htm

"Hawker, Harry George (1889-1921)" Sheely, Thomas, Australian Dictionary of Biography, accessed 2012, http://adb.anu.edu.au/biography/hawker-harry-george-6605.

"Lawrence Sperry: Autopilot Inventor and Aviation Innovator," Weider History Group, accessed 2012, www.HistoryNet.com.

"Maurice Ducrocq," Grace's Guide – The Best of British Engineering 1750 – 1960s, accessed 2012, www.gracesguide.co.uk /Maurice_Ducrocq.

"Raynham Family History" accessed 2012, www.rayham.org.

"Sunbeam Cars," Wolverhampton Museum of Industry, accessed 2012, http://www.historywebsite.co.uk/Museum/Transport/Cars/Sunbeam.htm

"Sunbeam Side Valve Engines: Crusader" Wolverhampton Museum of Industry, accessed 2012.

"The Era of the Dirigible," *U.S. Centennial of Flight Commission*, accessed 2012, www.centennialofflight.gov.

"The St. Annes Story," accessed 2012, www.lythamstannes.com/stannes/town_history.htm.

"Thomas Sopwith," Wikipedia, accessed 2012, https://en.wikipedia.org/wiki/Thomas_Sopwith.

"Vickers Vimy," Wikipedia, accessed 2012, https://en.wikipedia.org/wiki/Vickers_Vimy

About The Author

JOSEPH F ALCOCK is a former aerospace engineer. He was employed for more than forty years by Pratt & Whitney Rocketdyne (PWR), a division of United Technologies and was a project engineer on many of the company's commercial and military jet engines and rocket engines. Before he retired, he was the pre-launch operations manager for the booster engines in the Atlas III and Atlas V launch vehicles. He has written many proposals for PWR and several articles, stories and instruction manuals for local organizations since he retired in 2006.

Joe enjoys golf, boating, movies, and history. He discovered Captain Jack Alcock's early life and accomplishments, not covered in detail in other writings, were just as fascinating as his famous non-stop flight across the Atlantic with navigator Lieutenant Arthur Brown. Joe decided to give readers a front-row seat to the birth of aviation as told through Jack's life as a pioneer aviator.

End Notes

Preface

[1] KBE, Knight Commander of the Order of the British Empire; DSC, Distinguished Service Cross; RAF, Royal Air Force.

Acknowledgments

[1] MBE, Member of the Order of the British Empire; RAF, Royal Air Force.

Chapter 1

[1] Jack was born in Manchester 6 November 1892 and attended St. Thomas' Primary School in Stockport from 1898 to 1900. When his father took a job as a chauffeur for the owner of a brewery in Blackpool in 1900, Jack was enrolled in Heyhouses at St. Annes on Sea until his father moved back to Manchester in 1905 to become the circulation manager of Kemsley House Newspapers.
[2] Parkin is a sweet and sticky gingerbread cake eaten in autumn, particularly in Yorkshire.

Chapter 2

[1] Truck.
[2] Machine was the common term for an aeroplane in the early years.
[3] Empty weight.

Chapter 3

[1] Scarves were used by pilots to keep their necks warm in the open cockpits of that day. High-collared coats kept them warm, but restricted their head movement, which was absolutely necessary in wartime. In the Great War, the color white added to the debonair image the pilots wanted to create.
[2] After Royce died in 1933, the second 'R' in the Rolls-Royce emblem was changed from red to black.

Chapter 5

[1] Rotary air-cooled engines had a fixed crankshaft with the cylinders rotating around the crankshaft to provide adequate cooling of the cylinders. They also had a weight advantage as the rotation cylinders served as a flywheel as well. Later, radial air-cooled engines, with fixed cylinders and rotating crankshafts would replace the rotaries, but only after airspeeds were sufficient to cool the stationary cylinders.
[2] A pusher design had the engine mounted behind the wings; a tractor design in front of the wings.

Chapter 6

[1] A direct quote of Maurice Ducrocq.
[2] Flight manoeuvres at two-hundred feet are considered very dangerous due to the possibility of a stall and no time to recover from such a low altitude.

Chapter 7

[1] Both Henri and Maurice Farman built flying machines, but of different designs. Maurice's was generally known as the better machine due to its superior low speed performance.

Chapter 8

[1] There was at least one other project to cross the Atlantic in 1914 in response to the *Daily Mail* prize. A Curtis flying boat, *America*, to be flown by John Porte. The beginning of hostilities caused this attempt to be cancelled.

Chapter 11

[1] A loop-the-loop is a vertical circle where the aircraft is fully inverted at the top of the loop.

Chapter 12

[1] Now Istanbul.
[2] The Sopwith Camel, being a light aircraft, had strong gyroscopic forces from the engine and propeller. These forces caused the pilots to favor right turns in combat situations.
[3] First developed by Fokker for the Germans, a gun synchroniser is attached to the armament of an aircraft with a front mounted engine so that it can fire through the arc of a spinning propeller without the bullets striking the blades. The concept gave the Germans a distinct advantage in aerial combat and was

known as the Fokker Scourge when it was introduced in 1915. The Allies quickly adopted the same technology.
[4] The Sopwith Camel was an outstanding fighter and was credited with shooting down 1,294 enemy aircraft, more than any other Allied fighter of the Great War. It was also used as a ground-attack aircraft, especially near the end of the war, when newer fighters could out-perform it.

Chapter 13

[1] In Ottoman Turkey in the late 19th century and until The Great War, Bey was the honourary equivalent of the English speaking address Sir and followed the name. Thus, the English equivalent of Djavid Bey would be David Sir.
[2] Now the Gezde River, which is located in southwestern Turkey.
[3] The stage of a theatre, located between the curtain and the orchestra.
[4] Turkish rest-house.
[5] Treacle (sweetener) made from raisins.

Chapter 14

[1] Edward Samson Alcock later became a famous British aviator.
[2] Known as Indicated Air Speed.
[3] True Air Speed is approximately 2% higher per 1,000 feet of altitude, than the reading on the air speed metre for the altitudes and air speeds of 1919 aircraft. There is also a correction for instrument error. Thus, True Air Speed = (Calibrated Air Speed) x (Altitude Correction) and Calibrated Air Speed = (Indicated Air Speed) x (Instrument Correction).
[4] The angle created by the two sightings is constant regardless of altitude. The distance traveled between sightings is the altitude multiplied by a constant (the constant is related to the number of radians traveled). This distance, divided by the time, is the surface speed.
[5] Odd, meaning occasional.
[6] A London show.

Chapter 15

[1] From the 8th of May to the 31st of May 1919, a U.S. Navy Curtiss NC-4 flying boat, under the command of Albert Read, flew 4,526 statute miles from New York to Plymouth, England via Newfoundland, the Azores and Portugal in 53 hours, 58 minutes spread over 23 days, but didn't qualify for the *Daily Mail's* prize, because of several stops they made en route.

Chapter 16

[1] The propeller was actually a wind turbine (or windmill) that was mounted on the wing to drive a generator, which in turn supplied electrical power to the wireless.
[2] In 1937, the RAF defined the "Basic Six," the six instruments a pilot would need, as an absolute minimum, to fly an airplane. They were: 1) airspeed indicator, 2) attitude indicator (invisible horizon), 3) turn and bank indicator, 4) altimetre (altitude metre), 5) compass, and 6) vertical speed indicator. Alcock and Brown had all of these instruments except possibly for the vertical speed indicator. They also had a sextant and an instrument clock. Brown had fitted a spirit level to his sextant in case he could not see the horizon when taking a shot.
[3] A legend was later created that Brown walked out onto the wing to clear the snow from the petrol gauge. However, it is generally considered just a legend by the Alcock family and other experts in England. Brown stood up in the cockpit and was able to clear the snow off the fuel gauge without going out on the wings.
[4] Altimetre.
[5] Later Alcock cabled the Royal Aero Club, "Landed Clifden 8:40 a.m. (Greenwich Mean Time) June 15, Vicker Vimy machine leaving Newfoundland 4:28 p.m. (Greenwich Mean Time) June 14. Total time, 16 hours, 12 minutes. Signed Captain Alcock (Pilot), Lieutenant Brown (Navigator)."

Chapter 17

[1] Now Dun Laoghaire.
[2] Technically speaking, it was possible for a successful flight across Atlantic in 1917. The Germans designed and built the Poll Giant to bomb New York, and the Royal Flying Corps proposed to make the flight for propaganda reasons. The Handley Page O/400 could have been modified to carry enough fuel, as could German aircraft, and other British designs such as the DH10, or Felixstowe flying boats.

Chapter 18

[1] A borough in greater London on the west side.
[2] Took souvenirs from the machine.

Epilogue

[1] A sleeping pill. Also, there is continuing speculation on how accidental the overdose was.
[2] MBE, Member of the Order of the British Empire.

Made in the USA
Monee, IL
04 November 2026

67378678R00079